# Radiation, Isotopes, and Bone

**AMERICAN INSTITUTE OF BIOLOGICAL SCIENCES**

**and**

**U. S. ATOMIC ENERGY COMMISSION**

---

MONOGRAPH SERIES ON

## RADIATION BIOLOGY

FRANK FREMONT-SMITH, *Series Director*
AMERICAN INSTITUTE OF BIOLOGICAL SCIENCES

## ADVISORY COMMITTEE

AUSTIN M. BRUES, *Argonne National Laboratory*
LEO K. BUSTAD, *Hanford Laboratory*
CYRIL L. COMAR, *Cornell University*
ERNEST C. POLLARD, *Pennsylvania State University*
CHARLES W. SHILLING, *Biological Science Communications Project*

## MONOGRAPH TITLES AND AUTHORS

RADIATION, RADIOACTIVITY, AND INSECTS
R. D. O'BRIEN, *Cornell University*
L. S. WOLFE, *Montreal Neurological Institute*

RADIATION, ISOTOPES, AND BONE
F. C. MCLEAN, *University of Chicago*
A. M. BUDY, *University of Chicago*

RADIATION AND THE IMMUNE MECHANISMS (in press)
W. H. TALIAFERRO, *Argonne National Laboratory*
L. G. TALIAFERRO, *Argonne National Laboratory*
B. N. JARASLOW, *Argonne National Laboratory*

(IN PREPARATION)

HUMAN RADIOBIOLOGY
V. P. BOND, *Brookhaven National Laboratory*
T. M. FLIEDNER, *Brookhaven National Laboratory*
J. D. ARCHAMBEAU, *Brookhaven National Laboratory*

TRANSPLANT IMMUNITY AND RADIATION
J. F. LOUTIT, *Radiobiological Research Unit, Harwell*
H. S. MICKLEM, *Pasteur Institute*

PHYSICAL ASPECTS OF RADIOISOTOPES IN THE HUMAN BODY
F. W. SPIERS, *University of Leeds*

LIGHT: PHYSICAL AND BIOLOGICAL
W. D. MCELROY, *Johns Hopkins University*
H. H. SELIGER, *Johns Hopkins University*

SPACE RADIATION BIOLOGY
C. A. TOBIAS, *Donner Laboratory*
P. TODD, *Donner Laboratory*

MUTAGENESIS
I. I. OSTER, *Institute for Cancer Research*

# Radiation, Isotopes, and Bone

Franklin C. McLEAN
*Department of Physiology*
*University of Chicago*
*Chicago, Illinois*

Ann M. BUDY
*Department of Physiology*
*University of Chicago*
*Chicago, Illinois*

Prepared under the auspices of the Division of Technical Information, United States Atomic Energy Commission

1964

ACADEMIC PRESS • New York and London

ACADEMIC PRESS INC.
111 Fifth Avenue, New York, New York 10003

*United Kingdom Edition published by*
ACADEMIC PRESS INC. (LONDON) LTD.
Berkeley Square House, London W.1

Library of Congress Catalog Card Number: 63-23199

PRINTED IN THE UNITED STATES OF AMERICA

# FOREWORD

This monograph is one in a series developed through the cooperative efforts of the American Institute of Biological Sciences and the U. S. Atomic Energy Commission's Division of Technical Information. The goal in this undertaking has been to direct attention to biologists' increasing utilization of radiation and radioisotopes. Their importance as tools for studying living systems cannot be overestimated. Indeed, their application by biologists has an added significance, representing as it does the new, closer association between the physical and biological sciences.

The association places stringent demands on both disciplines: Each must seek to understand the methods, systems, and philosophies of the other science if radiation biology is to fulfill its promise of great contributions to our knowledge of both the normal and the abnormal organism. Hopefully, the information contained in each publication will guide students and scientists to areas where further research is indicated.

The American Institute of Biological Sciences is most pleased to have had a part in developing this Monograph Series.

JOHN R. OLIVE
Executive Director
American Institute of Biological Sciences

*October 1963*

# PREFACE

Of all the tissues of the body, none has a greater attraction for a variety of radioisotopes, nor holds them more firmly nor for a longer period of time, than does bone. Because of their predilection for the skeleton, many of the isotopes have come to be known as "bone-seekers." Not all of these, however, react with bone in the same manner. A few—more particularly some of the alkaline earths, including radium, strontium, and several of the isotopes of calcium—may substitute for stable calcium ($Ca^{40}$) in the mineral of bone, and once deposited there may remain fixed indefinitely. Others, such as plutonium, yttrium, and thorium, are deposited in the bone matrix rather than in the mineral, but they also may remain in the skeleton for long periods of time.

The properties that lead radioisotopes to seek bone, and to remain deposited therein, have increased both the hazards of radiation and the opportunities to study its effects. This book deals with both topics, and in order to provide a background for discussion, it has been found desirable to include a considerable amount of material concerning the physiology and biochemistry of skeletal tissue. The pathology of the effects of radiation, as well as the use of tracer amounts of radioisotopes in the study of the pathological physiology of bone, are given consideration, although not exhaustively.

There are two important landmarks in the history of the study of radioactive elements in relation to bone. The first stage was the recognition of the toxic and carcinogenic effects of ingested radium, participated in early by Blum (1924), Hoffman (1925), and Martland and Humphries (1929). The story is still incomplete, but the tragedy of the radium dial painters has been in review now for four decades. Another epoch was opened by Chiewitz and Hevesy in 1935, when they found, by the use of radiophosphorus ($P^{32}$), that there is a rapid exchange of phosphate between plasma and bone. The exchange reactions with the skeleton and the kinetics of the bone mineral still provide fascinating fields for study.

These landmarks were recorded before the advent of the cyclotron and of nuclear reactors greatly increased the availability of radioisotopes. While no doubt these developments would have followed in any case, but at a slower rate, the work incident to the exploitation of nuclear fission, during World War II, gave them added impetus and greatly accelerated progress in the study

of the atomic nucleus. Concurrently, major advances were made in the study of biological effects of radiation, and thus the discipline of radiobiology emerged.

This volume, then, is concerned with a special field of radiobiology, i.e., radiation, in all its aspects in relation to bone. The first symposium devoted to this particular field was organized by the Council for International Organizations of Medical Sciences (CIOMS) and was held at Princeton, New Jersey, August 29 to September 2, 1960. The symposium was under the leadership of Professor Pierre Lacroix, himself a pioneer in the application of radioisotopes to the study of bone, and was published under the title, "Radioisotopes and Bone," by Blackwell Scientific Publications, Oxford, in 1962. It is an important source for material on its subject, and the authors have drawn freely on this source in compiling the subject matter of the present volume.

The authors wish also to acknowledge with gratitude the assistance given by representatives of the American Institute of Biological Sciences (AIBS) and of the U.S. Atomic Energy Commission (AEC), who not only sponsored this volume but aided at every step in its production. While we must accept sole responsibility for the book and for its shortcomings, errors, and omissions, we could not have arrived at this point without the help of many individuals, too numerous to mention by name.

*Chicago, Illinois*
*October, 1963*

FRANKLIN C. MCLEAN
ANN M. BUDY

# CONTENTS


*Foreword* ........ v

*Preface* ........ vii

**CHAPTER 1 • Introduction** ........ 1

1. *Atomic and Nuclear Structure* ........ 2
2. *Radioactivity* ........ 7
3. *Radioactive Decay* ........ 9
4. *Radioactive Nuclides* ........ 10
5. *Radioactive Tracers* ........ 11

*References* ........ 11

**CHAPTER 2 • The Histophysiology of Bones and Teeth** 13

1. *Bone as a Tissue* ........ 13
2. *Formation and Organization of Bone* ........ 18
3. *The Mineral of Bone* ........ 22
4. *Teeth and Supporting Tissues* ........ 23

*References* ........ 29

**CHAPTER 3 • Bone-Seeking Elements** ........ 31

1. *Introductory Statement* ........ 31
2. *Periodic Table of Elements* ........ 31
3. *Transmutation and Translocation* ........ 33

*References* ........ 45

**CHAPTER 4 • Detection and Measurement of Radioactivity** 47

1. *Ionization and Excitation* ........ 47
2. *Nuclear Radiation Detectors* ........ 48
3. *Pulse Systems* ........ 48
4. *Nonpulse Systems* ........ 59

*References* ........ 60

**CHAPTER 5 • The Mineral of Bones and Teeth** .......... 61

1. *Crystal Structure and Chemical Composition of the Bone Mineral* .......... 61
2. *Crystals of Bone Mineral* .......... 62
3. *Dynamics of the Bone Mineral* .......... 71
4. *Surface Chemistry* .......... 74

*References* .......... 75

**CHAPTER 6 • Distribution of Radioelements in Bone** .... 79

1. *Carbonate in Bone* .......... 79
2. *The Sodium of Bone* .......... 79
3. *Magnesium in Bone* .......... 81
4. *Zinc in Bone* .......... 81
5. *Fluoride in Bone* .......... 84
6. *Other Trace Elements and Bone* .......... 85
7. *Radium and Plutonium in Bone* .......... 88

*References* .......... 88

**CHAPTER 7 • Kinetics of the Bone Mineral** .......... 91

1. *Introductory Statement* .......... 91
2. *Kinetic Studies—Interpretation* .......... 92
3. *Whole-Body Counting* .......... 98
4. *Compartmental Analysis* .......... 99
5. *The Power Function* .......... 100

*References* .......... 102

**CHAPTER 8 • Mineral Metabolism** .......... 105

1. *The Internal Environment of Bone* .......... 105
2. *Calcium of the Fluids of the Body* .......... 106
3. *Phosphates of the Fluids of the Body* .......... 108
4. *Physiological Factors in Distribution of Ions* .......... 109
5. *Discrimination Factors* .......... 109
6. *Transfer of Ions across Biological Membranes* .......... 111
7. *Absorption and Excretion of Calcium* .......... 112
8. *Differential Excretion* .......... 114
9. *Differential Secretion* .......... 114

*References* .......... 116

**CHAPTER 9 • Homeostasis in Mineral Metabolism** ........ 119

1. *Homeostasis of $Ca^{2+}$ in the Fluids of the Body* ............ 119
2. *Cybernetic Aspects of Homeostasis of $Ca^{2+}$* ............... 120
3. *The Parathyroid Hormone(s)* .......................... 122
4. *Calcitonin* ........................................ 123
5. *The Dual Mechanism* ................................ 123
6. *Mode of Action of the Parathyroid Hormone on Bone* ...... 124
7. *Use of Isotopes in the Study of Parathyroid Hormone* ...... 125
8. *Vitamin D* ........................................ 126

*References* ........................................ 128

**CHAPTER 10 • Application of Tracer Techniques to Bone** 131

1. *Introductory Statement* ............................. 131
2. *Autoradiography* .................................. 132
3. *Radioassay by Counting Methods* ..................... 138
4. *Radioisotope Scanning of Bone* ...................... 139
5. *Whole-Body Counting* .............................. 141

*References* ........................................ 144

**CHAPTER 11 • Radioisotopes and the Organic Portion of Bone** ........................................ 147

1. *Introductory Statement* ............................. 147
2. *The Matrix and Surfaces of Bone* ..................... 147
3. *Labeled Organic Molecules* .......................... 153

*References* ........................................ 157

**CHAPTER 12 • Effects of Radiation upon Bones** .......... 159

1. *Units of Radioactivity and of Radiation* ............... 159
2. *External Radiation* ................................. 160
3. *Internal Radiation* .................................. 161
4. *Radiation and Bone Tumors* .......................... 167

*References* ........................................ 169

**CHAPTER 13 • Natural Radioactivity and Radioactive Fall-out** ........................................ 171

1. *Natural Radioactivity* ............................... 171
2. *Naturally Occurring Radioactive Elements* ............... 172
3. *Radioactive Fallout* ................................. 177

4. *Strontium-90* ........ 179
5. *Removal of Radioisotopes from Bone* ........ 183
6. *Whole-Body Exposure* ........ 183
*References* ........ 184

**CHAPTER 14 • The Pathology and Pathological Physiology of Bone** ........ 187

1. *Introductory Statement* ........ 187
2. *Kinetic Methods* ........ 187
3. *Kinetic Studies—Results* ........ 189
4. *Localization by External Counting* ........ 189
5. *Whole-Body Counting* ........ 190
6. *Clinical Interpretations* ........ 190
*References* ........ 197

AUTHOR INDEX ........ 199

SUBJECT INDEX ........ 205

# ***Radiation, Isotopes, and Bone***

# Chapter 1

# *Introduction*

Within the past few decades and especially during and since World War II, a new field dealing with the biological effects of radiation and of radioactive substances, including those upon bone, has been opened for exploration. That certain radioactive elements have a special affinity for bone was first appreciated by Blum,[1] who in 1924 reported on the osteomyelitis of the mandible and maxilla which occurred in dial painters using luminous paint, and by Hoffman,[2] who in 1925 attributed necrosis of the jaw bones to radium or mesothorium. In 1929, Martland and Humphries[3] described the occurrence—also in dial painters—of osteogenic sarcoma.

The first use of an artificial radioelement as a tool in animal experimentation was reported by Chiewitz and Hevesy[4] in 1935; they found by the use of radiophosphorus ($P^{32}$) that there is a rapid exchange of phosphate between plasma and bone. The introduction of the cyclotron, the subsequent development of nuclear reactors, the exploitation of these and other instruments for the production of radioisotopes and even of new elements, and the testing of weapons dependent upon nuclear fission have increased both the hazards of radiation and the opportunities for making use of its properties in biological investigation.

Major advances have been made in the study of the biological effects of radiation, and the discipline of radiobiology has thus emerged. In this monograph we are concerned with radiation biology mainly insofar as bones and teeth are influenced by ionizing radiation. Ionizing radiation itself and the metabolism of radioactive isotopes have little place in normal physiology and this limits our treatment of the effects of radiation, external and internal, upon bones and teeth in a large degree to its pathophysiology. Incidental to this, but of importance to the organism as a whole, is the metabolism of elements (including both their stable and radioactive isotopes) having a special affinity for the skeletal system. These elements must be considered on the basis of their absorption, transport, retention, and excretion; in all of these processes the circulating fluids of the body and the soft tissues are also of moment. Our treatment of these topics will be limited to the bone-seeking elements; those elements of relatively little importance to the skeletal system will receive less attention in this volume.

An important by-product of the advances in radiobiology has been a

rapid increase in the usefulness of radioisotopes as *tracers* for the study of physiological, biochemical, and pathological processes in the organism; this implies that the isotopes are used in quantities small enough to limit the amount of radioactivity to levels which will insure minimal or negligible biological effects. Much of this volume will be devoted to the use of tracer quantities of radioelements in elucidating the histophysiology and histopathology of bones and teeth. While the use of radioelements in living organisms is attended by certain limitations, their introduction into studies of this type has marked an important step forward. Thus, the analysis of physiological mechanisms with tracers has come into extensive use. Metabolic studies of the elements common to bones and teeth, notably calcium and phosphorus and even the rare elements, have been facilitated. Information not otherwise obtainable has appeared in the literature, and formulation of new concepts has been made possible.

It will become apparent in this volume that, even though the use of radioisotopes has added a powerful tool to the armamentarium for the study of bone, this is not enough for clarification of the numerous problems in this field. Important as they are in the study of the physiology and pathology of bone and in the clinic, the isotopes are used to the best advantage when combined with other tools and other methods—some new, some well established. For this reason, it has not seemed advisable to restrict this monograph to isotopic methods; instead, the effort has been to show how their use may supplement the many other approaches to the topics to be discussed.

It is not possible to treat the use of isotopes in the study of bones and teeth without some consideration of the physiology and biochemistry and, to some extent, the pathological physiology of the skeletal and dental systems. The intent is to include enough of the biological principles subject to exploration by isotopic methods to supply a background for their use. Moreover, the biological application of radioactive elements depends in large part upon the advances made in nuclear physics. Certain aspects of these topics are essential for proper orientation in the subject matter herein presented.

## 1. Atomic and Nuclear Structure

The subdivision of matter into small particles called *atoms*, advanced by John Dalton as a hypothesis in 1805, is now accepted as a fact. It is, however, well established that the atom itself consists of even smaller particles, common to all atoms, and that the characteristics that distinguish the atom of one element from that of another depend upon the number and arrangement of these elementary particles within the atom.

Each atom consists of one *nucleus* and one or more *electrons*. The nucleus is a small, positively charged heavy particle around which the electrons move in orbit. The nucleus contains almost the entire weight of the atom; the electrons are of extremely light weight, and each one carries the unit negative electronic charge *e*. The nucleus carries positive charges, each equal in magnitude to the charge of one electron. The number of positive charges carried by the nucleus is equal to the number of electrons present in the same atom.

The electrons are in continuous motion, are at varying distances from the nucleus, and reside in layers or shells centered around the nucleus. Only the

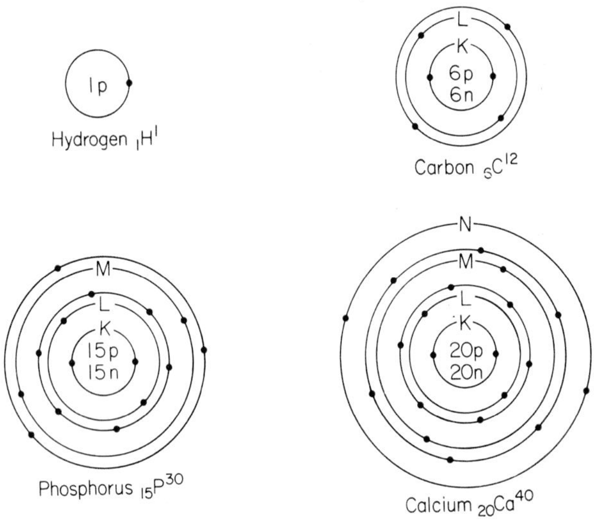


FIG. 1.1. Schematic representation of H, C, P, and Ca atoms indicating the number of neutrons (n) and protons (p) within the nucleus. The orbital electrons are represented by dots in the *K*, *L*, *M*, *N* shells and subshells.

outer shell of the atom is involved in chemical reactions. These loosely attached peripheral electrons in the outermost shell play the dominant part in determining the physical and chemical properties of the atom. The number of shells and subshells for any particular atom characterizes the element while the number of orbital electrons determines the place at which an element appears in the periodic table.

The atomic nucleus is an aggregate of *nucleons*, or nuclear constituents, of which there are two varieties: *protons* (p) and *neutrons* (n). Either variety serves as the unit of mass, and to each is assigned mass number 1. The important difference between them is that each proton carries a unit positive charge, equal in magnitude to the negative electronic charge *e*, while the neutron carries no charge (Fig. 1.1).

TABLE 1.1

TABLE OF INTERNATIONAL ATOMIC WEIGHTS

| Element | Symbol | Atomic number | Atomic weight based on oxygen-16 1959[a] | Atomic weight based on carbon-12 1961[b] |
|---|---|---|---|---|
| Actinium | Ac | 89 | [227][c] | — |
| Aluminum | Al | 13 | 26.98 | 26.9815 |
| Americium | Am | 95 | [243] | — |
| Antimony | Sb | 51 | 121.76 | 121.75 |
| Argon | Ar | 18 | 39.944 | 39.948 |
| Arsenic | As | 33 | 74.92 | 74.9216 |
| Astatine | At | 85 | [210] | — |
| Barium | Ba | 56 | 137.36 | 137.34 |
| Berkelium | Bk | 97 | [249] | — |
| Beryllium | Be | 4 | 9.013 | 9.0122 |
| Bismuth | Bi | 83 | 208.99 | 208.980 |
| Boron | B | 5 | 10.82 | 10.811[d] |
| Bromine | Br | 35 | 79.916 | 79.909[e] |
| Cadmium | Cd | 48 | 112.41 | 112.40 |
| Calcium | Ca | 20 | 40.08 | 40.08 |
| Californium | Cf | 98 | [251] | — |
| Carbon | C | 6 | 12.011 | 12.01115[d] |
| Cerium | Ce | 58 | 140.13 | 140.12 |
| Cesium | Cs | 55 | 132.91 | 132.905 |
| Chlorine | Cl | 17 | 35.457 | 35.453[e] |
| Chromium | Cr | 24 | 52.01 | 51.996[e] |
| Cobalt | Co | 27 | 58.94 | 58.9332 |
| Copper | Cu | 29 | 63.54 | 63.54 |
| Curium | Cm | 96 | [247] | — |
| Dysprosium | Dy | 66 | 162.51 | 162.50 |
| Einsteinium | Es | 99 | [254] | — |
| Erbium | Er | 68 | 167.27 | 167.26 |
| Europium | Eu | 63 | 152.0 | 151.96 |
| Fermium | Fm | 100 | [253] | — |
| Fluorine | F | 9 | 19.00 | 18.9984 |
| Francium | Fr | 87 | [223] | — |
| Gadolinium | Gd | 64 | 157.26 | 157.25 |
| Gallium | Ga | 31 | 69.72 | 69.72 |
| Germanium | Ge | 32 | 72.60 | 72.59 |
| Gold | Au | 79 | 197.0 | 196.967 |
| Hafnium | Hf | 72 | 178.50 | 178.49 |
| Helium | He | 2 | 4.003 | 4.0026 |
| Holmium | Ho | 67 | 164.94 | 164.930 |
| Hydrogen | H | 1 | 1.0080 | 1.00797[d] |
| Indium | In | 49 | 114.82 | 114.82 |
| Iodine | I | 53 | 126.91 | 126.9044 |
| Iridium | Ir | 77 | 192.2 | 192.2 |

TABLE 1.1 (*Continued*)

| Element | Symbol | Atomic number | Atomic weight based on oxygen-16 1959[a] | Atomic weight based on carbon-12 1961[b] |
|---|---|---|---|---|
| Iron | Fe | 26 | 55.85 | 55.847[c] |
| Krypton | Kr | 36 | 83.80 | 83.80 |
| Lanthanum | La | 57 | 138.92 | 138.91 |
| Lead | Pb | 82 | 207.21 | 207.19 |
| Lithium | Li | 3 | 6.940 | 6.939 |
| Lutetium | Lu | 71 | 174.99 | 174.97 |
| Magnesium | Mg | 12 | 24.32 | 24.312 |
| Manganese | Mn | 25 | 54.94 | 54.9380 |
| Mendelevium | Md | 101 | [256] | — |
| Mercury | Hg | 80 | 200.61 | 200.59 |
| Molybdenum | Mo | 42 | 95.95 | 95.94 |
| Neodymium | Nd | 60 | 144.27 | 144.24 |
| Neon | Ne | 10 | 20.183 | 20.183 |
| Neptunium | Np | 93 | [237] | — |
| Nickel | Ni | 28 | 58.71 | 58.71 |
| Niobium | Nb | 41 | 92.91 | 92.906 |
| Nitrogen | N | 7 | 14.008 | 14.0067 |
| Nobelium | No | 102 | [254] | — |
| Osmium | Os | 76 | 190.2 | 190.2 |
| Oxygen | O | 8 | 16.0000 | 15.9994[d] |
| Palladium | Pd | 46 | 106.4 | 106.4 |
| Phosphorus | P | 15 | 30.975 | 30.9738 |
| Platinum | Pt | 78 | 195.09 | 195.09 |
| Plutonium | Pu | 94 | [242] | — |
| Polonium | Po | 84 | 210 | — |
| Potassium | K | 19 | 39.100 | 39.102 |
| Praseodymium | Pr | 59 | 140.92 | 140.907 |
| Promethium | Pm | 61 | [147] | — |
| Protactinium | Pa | 91 | 231 | — |
| Radium | Ra | 88 | 226 | — |
| Radon | Rn | 86 | 222 | — |
| Rhenium | Re | 75 | 186.22 | 186.2 |
| Rhodium | Rh | 45 | 102.91 | 102.905 |
| Rubidium | Rb | 37 | 85.48 | 85.47 |
| Ruthenium | Ru | 44 | 101.1 | 101.07 |
| Samarium | Sm | 62 | 150.35 | 150.35 |
| Scandium | Sc | 21 | 44.96 | 44.956 |
| Selenium | Se | 34 | 78.96 | 78.96 |
| Silicon | Si | 14 | 28.09 | 28.086[d] |
| Silver | Ag | 47 | 107.873 | 107.870[e] |
| Sodium | Na | 11 | 22.991 | 22.9898 |
| Strontium | Sr | 38 | 87.63 | 87.62 |
| Sulfur | S | 16 | 32.066 | 32.064[d] |

TABLE 1.1 (*Continued*)

| Element | Symbol | Atomic number | Atomic weight based on oxygen-16 1959[a] | Atomic weight based on carbon-12 1961[b] |
|---|---|---|---|---|
| Tantalum | Ta | 73 | 180.95 | 180.948 |
| Technetium | Tc | 43 | [99] | — |
| Tellurium | Te | 52 | 127.61 | 127.60 |
| Terbium | Tb | 65 | 158.93 | 158.924 |
| Thallium | Tl | 81 | 204.39 | 204.37 |
| Thorium | Th | 90 | [232] | 232.038 |
| Thulium | Tm | 69 | 168.94 | 168.934 |
| Tin | Sn | 50 | 118.70 | 118.69 |
| Titanium | Ti | 22 | 47.90 | 47.90 |
| Tungsten | W | 74 | 183.86 | 183.85 |
| Uranium | U | 92 | 238.07 | 238.03 |
| Vanadium | V | 23 | 50.95 | 50.942 |
| Xenon | Xe | 54 | 131.30 | 131.30 |
| Ytterbium | Yb | 70 | 173.04 | 173.04 |
| Yttrium | Y | 39 | 88.91 | 88.905 |
| Zinc | Zn | 30 | 65.38 | 65.37 |
| Zirconium | Zr | 40 | 91.22 | 91.22 |

[a] From Hodgman, *Handbook of Chemistry and Physics.*[6]

[b] From *Report of the Commission on Atomic Weights.*[7]

[c] The values in brackets are the mass numbers of the most stable isotopes.

[d] The atomic weight varies because of natural variations in the isotopic composition of the element. The observed ranges are boron, ± 0.003; carbon, ± 0.00005; hydrogen, ± 0.00001; oxygen, ± 0.0001; silicon, ± 0.001; sulfur, ± 0.003.

[e] The atomic weight is believed to have an experimental uncertainty of the following magnitude: bromine, ± 0.002; chlorine, ± 0.001; chromium, ± 0.001; iron, ± 0.003; silver, ± 0.003. For other elements the last digit given is believed to be reliable to ± 0.5.

The proton is identical with the hydrogen ion, $H^+$, and forms the nucleus of the hydrogen atom. It has unit mass and a unit positive electrical charge. Every element in the periodic table has a different nuclear charge which corresponds to the number of protons in the nucleus; this is its $Z$ or *atomic number,* which identifies the element and the nuclear species. The atomic number of hydrogen is 1. The hydrogen molecule includes two hydrogen atoms, and therefore two protons; they share two electrons.

The neutron, the second constituent of the nucleus, adds to mass in proportion to the number present and thus contributes to the nuclear mass but not to its charge. The sum of the number of protons in a nucleus plus the number of neutrons is the $A$ or *mass number.* This number, together with the atomic number or nuclear charge, characterizes a member of the nuclear species, or *nuclide.*

An isotope is one of a nuclear species, or nuclide, which has the same number of protons in the nucleus, or the same $Z$ or atomic number, and hence the same nuclear charge. Isotopes of the same nuclide have different numbers of neutrons and consequently different $A$ or mass numbers. Since the mass number is the sum of the number of protons and of neutrons in the nucleus, different elements may have isotopes with the same mass number. For complete identification of an isotope, the convention includes a subscript for the atomic number, the chemical symbol of the element, and a superscript for the mass number. Thus ${}_{20}Ca^{40}$ represents the calcium isotope with 20 protons and 20 neutrons; the mass number is the sum of these. $Z$ and $A$ are always whole numbers.

As a result of the unification of the physical and chemical atomic weight scales, the mass of an atom of a nuclide, or the average mass of an atom in an element, is now expressed as the ratio of this mass to one-twelfth of the mass of a neutral atom of carbon-12. This ratio is the *relative nuclidic mass* of the nuclide or the *relative atomic mass* of the element.[5] For convenience, there are listed in Table 1.1 the name, symbol, and atomic number of the elements. Also included are the international atomic weights based on oxygen[6] and the revised unified weights based on carbon.[7]

This description of the structure of the atom and of its nucleus ignores many particles and antiparticles that are not essential to the subject matter of this monograph. The details of the internal structure of the nucleus are not known; it is certainly an oversimplification to regard it as a number of protons and neutrons packed closely together. There is evidence that protons and neutrons are themselves aggregates of smaller particles and also that these nucleons are bound together by particles known as pi mesons.

## 2. Radioactivity

Radioactivity depends upon the emission of energy in the form of *radiation* and results from a condition of instability in the nucleus which seeks a level of stability. The lighter elements are most stable if the numbers of protons and neutrons are approximately equal. The heavier nuclei have a smaller number of protons than of neutrons, and, with increasing atomic mass, a growing excess of neutrons occurs. Since each proton possesses a positive charge, the protons mutually repel each other, while the neutrons, having no charge, exhibit an attraction for each other and for protons. As the number of protons in the nucleus increases, more and more neutrons are required to separate them and to reduce their repulsive interaction. In a particular nuclide in which the number of neutrons relative to the number of protons is unusually large or small, the nucleus is *unstable;* it tends to return to a more

normal proton-neutron relation by emitting energy in either particulate or electromagnetic form, thereby transforming into a stable nucleus. The energy of any particle is measured in electron volts (ev) or multiples of this unit, i.e., kilo electron volts (kev) or million electron volts (Mev). An electron volt is the energy equal to the work done in transferring one electron against a potential difference of 1 volt.

Radiation is emitted from either the atom's unstable nucleus or its unstable orbital area. Emission of energy from a given nucleus is in one or more of three forms: (1) *alpha particles;* (2) *beta particles,* or electrons (either negatively or positively charged); and (3) *gamma rays,* or electromagnetic radiation. Alpha or beta radiation is associated with decay of the element, and results in spontaneous transmutation to another element with a different atomic or *Z* number.

Alpha particles carry two positive charges and are composed of two protons and two neutrons each; they are identical with the nucleus of the helium atom and their emission represents alpha decay, which occurs predominantly in elements with atomic numbers greater than 82. Emission of an alpha particle from the nucleus of an atom results in the loss of two positive charges and a reduction of 4 in mass number; the net effect is transmutation to the element with an atomic number lower by 2. Most of the heavy elements, with atomic numbers greater than 82, are unstable with respect to alpha decay.

Beta particles are high-speed electrons and may be either positively ($\beta^+$, positron) or negatively ($\beta^-$, negatron) charged. For a beta particle to be emitted from a nucleus which has none in its structure, an electron must be created in the act of its emission. This requires that either a neutron or a proton (commonly a proton) must undergo transformation to the other. A neutron may gain a positive charge, while releasing a negatively charged electron, and become a proton; conversely, a proton may lose its positive charge, release a positively charged electron, and become a neutron. At the same time in either case, there is released another particle, of infinitely small rest mass and without charge—the *neutrino.* When beta decay of a nucleus occurs, the mass number remains the same while the atomic number is either increased by 1 ($\beta^-$ decay) or reduced by 1 ($\beta^+$ decay). Beta decay thus results in transmutation to another element with an atomic number either less or greater by 1; formation of another isotope of the same element is not a possible consequence of beta decay.

Gamma rays consist of electromagnetic radiation emitted from the nucleus of an atom as *photons.* Except for differences in wavelength and, consequently, in penetrating power, gamma rays are identical with light rays; they have physical properties and hence biological effects similar to those of *X-rays.* The emission of gamma rays from the nucleus occurs as a concomitant of an

excited state of the nucleus. Ordinarily associated with alpha or beta decay or both, it represents an adjustment of the energy relations within the nucleus, and its terminal state is de-excitation or return to the ground state.

There are two situations in which only gamma rays are emitted from an atom; one of these results in transmutation of the element and may have bearing on the use of radioisotopes. This is the phenomenon of *electron capture* (EC), by which is meant that the nucleus captures an electron from the shell of the atom; this results in the transformation of one proton to a neutron and the reduction of the atomic number by one. An excited state of the nucleus may result, accompanied by the emission of gamma rays. Of the isotopes in common use, those that undergo transformation to other elements with emission of gamma rays only are: $Be^7$ to $Li^7$, $Sr^{85}$ to $Rb^{85}$, and $Ba^{133}$ to $Ce^{133}$. In addition, the transition of $F^{18}$ to $O^{18}$ and of $Na^{22}$ to $Ne^{22}$ occurs by a combination of $\beta^+$ decay and electron capture. The second phenomenon associated with gamma radiation alone is that of *isomeric transition* (IT), which represents a delay in adjusting to the ground state; this is commonly brief in duration and has not been shown to have physiological significance. Gamma radiation, while originating from isotopes usually undergoing alpha or beta decay, is not itself a part of decay.

## 3. Radioactive Decay

A radioactive substance disintegrates or decays in a manner characteristic of a first-order reaction, i.e., decay occurs in proportion to the number of radioactive atoms present. Because radioactive decay is a random process governed by a constant probability per unit time, the fraction of atoms in a very large sample decaying in a given time interval will be constant. This may be represented by

$$\frac{dN}{dt} = -\lambda N \qquad (1.1)$$

in which $N$ is the number of radioactive atoms present at any time $t$, $\lambda$ is the decay constant of a particular radioisotope, and $dN/dt$ is the rate of change of the number of radioactive atoms. The solution of this equation is the well-known exponential relation

$$N = N_0 e^{-\lambda t} \qquad (1.2)$$

in which $N_0$ is the number of radioactive atoms present at time $t = 0$. $N$ is here described as an exponential function of time. If log $N$ is plotted vs. $t$, the result is a straight line starting at $N = N_0$ for $t = 0$ with $N$ decreasing as $t$ increases. The half-life of the radioisotope is $0.693/\lambda$, i.e., the time it takes for $N$ to decrease to one-half its previous value no matter what time is

chosen as the starting point. Thus, the *physical half-life* of a radioactive element may be defined as the time in which each atom of a radioisotope has a 50% probability of decay. The net result is that during one physical half-life, a sample of radioisotope loses half of its activity.

## 4. Radioactive Nuclides

The term *nuclide* refers to any one of the forms a nucleus may assume; a nuclide is characterized by the atomic number of the element and the mass number of the particular isotope. The term isotope does not in itself imply radioactivity; an isotope may be either stable or radioactive (Fig. 1.2). A radioactive isotope may be designated as a *radioisotope* or *radionuclide*.

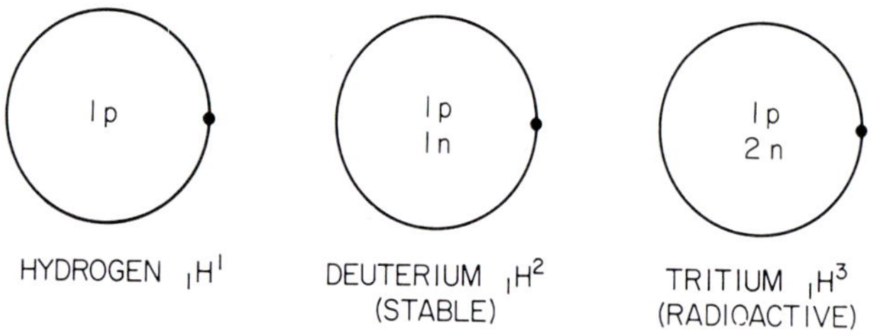


FIG. 1.2. Isotopes of hydrogen.

There are over 40 naturally occurring radioisotopes; others are produced artificially. The number of known isotopes of all the elements is of the order of 1500. New elements, as well as new isotopes, have been formed in the laboratory. In addition to the transmutations that occur naturally (Chapter 3), as in the decay of the heavy elements such as radium, transmutations may be induced. Of these events the most dramatic is nuclear fission, brought about in the atomic bomb or in the nuclear reactor, by fission of the atoms of plutonium or of uranium, and resulting in radioisotopes of the 34 elements in the periodic system from zinc (element 30) to the rare earth europium (element 63). In some instances more than one isotope of the same element may be formed as fission products. Other radioisotopes may be produced by bombardment of the nucleus with particles, more commonly neutrons, which add to the mass of the nucleus that absorbs them, resulting in a new isotope of the same element, with an increase by 1 in the mass number. Other examples of transmutation are *deuteron induced*, *alpha particle induced*, or *proton induced*. Neutron bombardment requires a source of *neutron flux* such as an atomic pile or nuclear reactor. Induction by deuterons, alpha particles, or protons requires acceleration of the particles, in order that they may acquire sufficient energy to enter the nucleus under bombardment. A variety of high-

energy accelerators is now available for research in particle physics, as well as for the formation of new isotopes.

## 5. Radioactive Tracers

The use of isotopes as tracers, preferably detectable in such minute amounts that their physiological effects are negligible, is based on the assumption that the radioisotope behaves chemically and in the organism in the same manner as the stable forms of the same element. $Ca^{45}$ and $Ca^{47}$, both radioactive, in tracer amounts reflect the behavior of the stable $Ca^{40}$. In those elements having low atomic weights, a difference in mass of an isotope may introduce a difference in behavior. Thus $H^2$ (deuterium) and $H^3$ (tritium) do not reflect accurately the behavior of stable hydrogen. To a lesser extent this may be true also of the isotopes of carbon and of nitrogen.

Tracer elements may be either native or foreign to the organism in which they are to be used. $Ca^{45}$ and $Ca^{47}$ serve to represent the stable $Ca^{40}$; $Sr^{85}$, $Sr^{89}$, $Sr^{90}$, and $Ra^{226}$ have proved useful in study of the fixation of calcium-like elements by bone. Strictly speaking, neither Sr nor Ra is foreign to the human organism, but the amounts of radium or of radioactive strontium normally present in the body are not detectable by methods in common use.

A further category of radioactive tracers consists of labeled compounds such as vitamins and hormones. The label is a radioactive atom, incorporated in the molecule; the distribution and metabolism, and to some extent the physiological effects, may be followed by observing the tracer atom. A variant of this category is the use of a labeled precursor which is subject to biosynthesis in the organism and results in the formation of a compound whose radioactivity renders it accessible to detection by radioassay and autoradiographic methods (Chapter 4).

### References

1. T. Blum, Osteomyelitis of the Mandible and Maxilla, *J. Am. Dental Assoc.*, **11**: 805, footnote (1924).
2. F. L. Hoffman, Radium (Mesothorium) Necrosis, *J. Am. Med. Assoc.*, **85**: 961-965 (1925).
3. H. S. Martland and R. E. Humphries, Osteogenic Sarcoma in Dial Painters Using Luminous Paint, *Arch. Pathol.*, **7**: 406-417 (1929).
4. O. Chiewitz and G. Hevesy, Radioactive Indicators in the Study of Phosphorus Metabolism in Rats, *Nature (London)*, **136**: 754-755 (1935).
5. E. A. Guggenheim, The Mole and Related Quantities, *J. Chem. Ed.*, **38**: 86-87 (1961).
6. C. D. Hodgman (Ed.), *Handbook of Chemistry and Physics*, 44th ed., The Chemical Rubber Publishing Co., Cleveland, 1962-63.
7. *Report of the Commission on Atomic Weights* (C. R. Vingt et Unième Conf. XXI, Montreal, 2-5 Août), pp. 281-322, Butterworths Scientific Publ., London, 1961.

## Bibliography

C. F. Behrens, *Atomic Medicine*, The Williams & Wilkins Co., Baltimore, 1959.

N. Bohr, *Atomic Physics and Human Knowledge*, John Wiley & Sons, Inc., New York, 1958.

T. F. Dougherty, W. S. S. Jee, C. W. Mays, and B. J. Stover (Eds.), *Some Aspects of Internal Irradiation*, Pergamon Press, New York, 1962.

R. D. Evans, *The Atomic Nucleus*, McGraw-Hill Book Co., New York, 1955.

O. Glasser (Ed.), *Medical Physics*, Vol. III, pp. 302-354, The Year Book Publishers, Inc., Chicago, 1960.

E. Grunwald and R. H. Johnsen, *Atoms, Molecules and Chemical Change*, Prentice-Hall, Inc., Englewood Cliffs, N. J., 1960.

G. Hevesy, *Radioactive Indicators*, Interscience Publishers, Inc., New York, 1948.

D. J. Hughes, *On Nuclear Energy*, Harvard University Press, Cambridge, 1957.

M. D. Kamen, *Isotopic Tracers in Biology*, 3rd ed., Academic Press Inc., New York, 1957.

A. Kolin, *Physics, Its Laws, Ideas, and Methods*, McGraw-Hill Book Co., Inc., New York, 1950.

E. H. Quimby, S. Feitelberg, and S. Silver, *Radioactive Isotopes in Clinical Practice*, Lea & Febiger, Philadelphia, 1958.

RCA Service Company, *Atomic Radiation*, Camden, N. J., 1957.

# Chapter 2

# *The Histophysiology of Bones and Teeth*

## 1. Bone as a Tissue

The characteristics of bone, and especially those that render it accessible to study by means of tracer elements, will be considered in greater detail in the chapters that follow. Certain features and certain concepts, however, are here treated briefly.

Bone is a highly specialized form of connective tissue, composed of branching cells in an intercellular substance and forming the skeleton or framework of the body of most vertebrates. There are certain characteristics that differentiate bone from other forms of connective tissue, of which there are many types, the most striking one being that it is hard. This hardness results from deposition, within a soft organic matrix, of a complex mineral substance, composed chiefly of calcium, phosphate, carbonate, and citrate. Bone has cells peculiar to it; they are specialized forms derived from cells common to connective tissue.

The interstitial substance, in addition to being calcified, has a fibrillar structure similar to that of connective tissue. Although the fibers are mainly those of collagen, reticular fibers are also found in bone. The ground substance, as in connective tissue, is characterized in bone by its content of mucopolysaccharides.[1-4] Connective tissue and bone have in common the important function of the support of organs or elements of organs. The mechanical and physiological functions of both tissues are interdependent.

### 1.1. The Cells of Bone

The cellular components of bone are associated with specific functions: *osteoblasts* with the formation of bone, *osteocytes* with the maintenance of bone as a living tissue, and *osteoclasts* with the resorption of bone. These cells, having common ancestors, are closely interrelated. During active growth, frequent transformations occur from one to another of the morphologically different forms while they retain the potencies common to all three. The morphology of a particular cell at any given time depends upon the function it is being called upon to perform. The transformations from one type to another, occurring spontaneously more frequently in developing bone, can be demonstrated in adult bone under certain conditions. The most striking examples of such responses are seen during the healing of

fractures,[5] in the rat under the influence of parathyroid extract,[6] or in the bird[7] or mouse[8] under the influence of estrogens.

The osteoblasts are located on the surface of a growing or developing bone, and they synthesize collagen. During active growth, they appear to be in a continuous layer and are frequently connected with one another by thin cytoplasmic processes. They are cuboidal in shape, with a breadth of

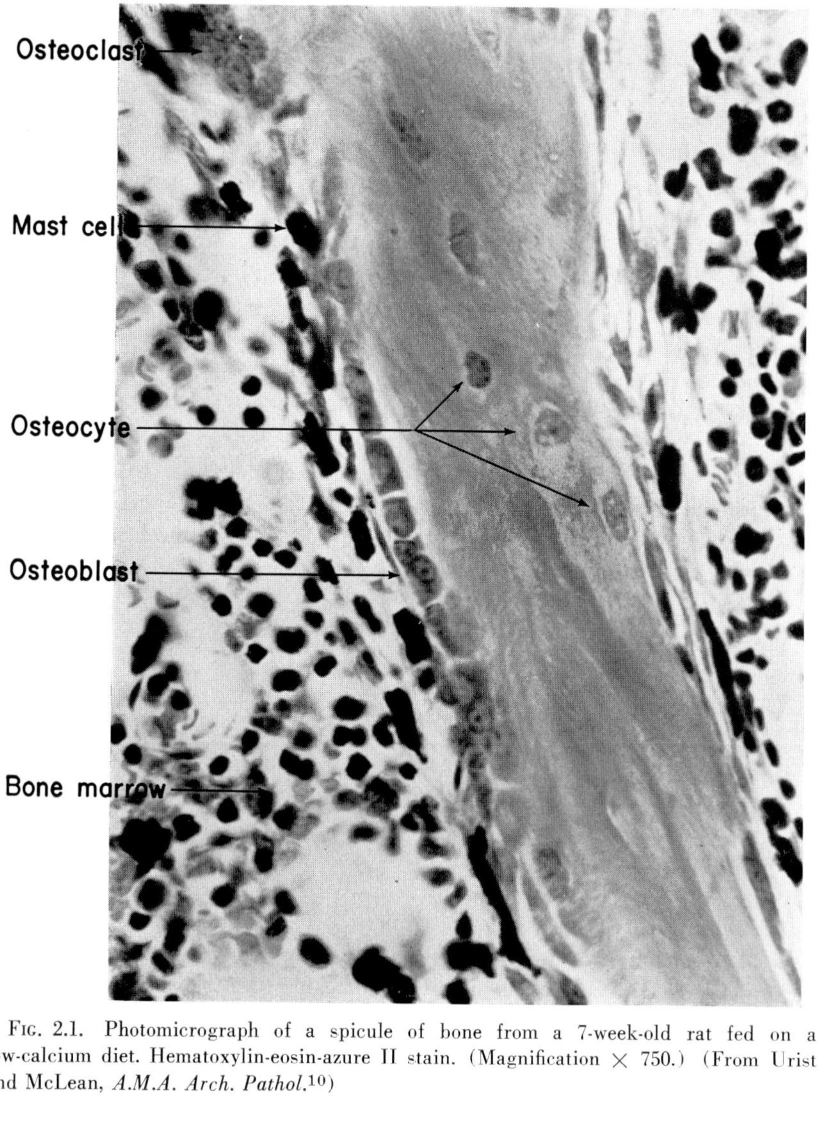


FIG. 2.1. Photomicrograph of a spicule of bone from a 7-week-old rat fed on a low-calcium diet. Hematoxylin-eosin-azure II stain. (Magnification × 750.) (From Urist and McLean, *A.M.A. Arch. Pathol.*[10])

15–20 μ. The typical morphological characteristics are absent or unrecognizable when the particular structure with which they have been associated is in a resting state; the osteoblasts then become spindle-shaped and resemble fibroblasts or reticular cells.

Osteocytes are osteoblasts that have been surrounded by their secretory products which in turn become calcified and form the interstitial substance. The cells are now enclosed within lacunae, and the cytoplasmic processes extend through apertures in the lacunae into canalicules in the bone, connecting the lacunae with each other and providing for the circulation of interstitial fluids.[9]

The osteoclast is a giant cell with a variable number of nuclei, often as many as 15 or 20. For many years, the osteoclasts have been considered to play the principal role in the resorption or dissolution of bone.

## 1.2. Interstitial Substance

The intercellular portion of bone is a calcified collagenous substance that makes up the great mass of bone. The interstitial substance includes the organic framework or matrix, the inorganic part or the mineral of bone, and water. The organic matrix has two chief components—the collagenous fibers and the ground substance. The organic fraction of compact bone may contribute as much as 35% of the dry, fat-free weight. Only a small part of this fraction consists of cells; the remainder, impregnated with the bone salt, is the bone matrix.

### *1.2.1. Collagen*

Bone collagen is a substance which yields glue or gelatin when boiled. It makes up 90–96% of the dry, fat-free weight of the organic fraction of bone. It occurs in fibrils, with double cross-banding at intervals averaging 640 A. It also gives a characteristic X-ray diffraction pattern. Chemically, it is characterized by a low content of aromatic amino acids and a high content of pyrrolidine amino acids and glycine. Hydroxyproline content is an index for the quantitative estimation of mature collagen, as is hexosamine for ground substance. Certain collagens, such as those from the tendons of the rat's tail and from the swim bladder of fish, are soluble in dilute acid and may be reprecipitated in fibrils, with typical periodic cross-banding, by addition of salts to the solution.[11] The individual fibrils, 0.3–0.5 μ in diameter, are often collected into small bundles 3–5 μ thick.[12] The role of collagen in the seeding or nucleation of crystals of the bone mineral will be explored in Chapter 5.

### *1.2.2. Ground Substance*

This is the extracellular and interfibrillar component of all connective tissues and is characterized chemically by its content of sulfated polysaccharides con-

taining hexosamines, or amino sugars.[1,2] Although small in amount, the ground substance is a physiologically important constituent of bone, possibly related to the calcification process.

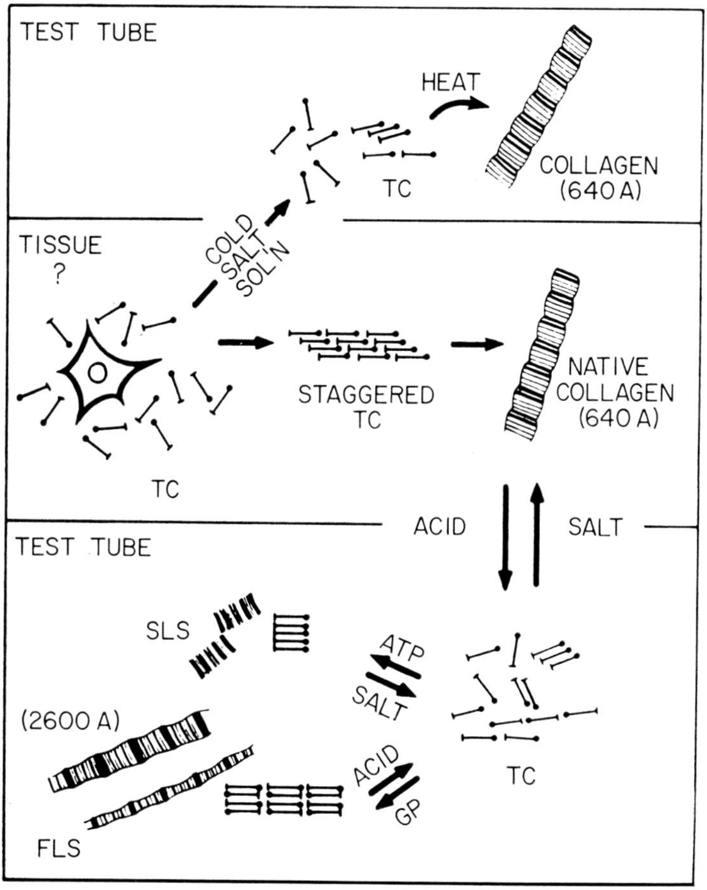


Fig. 2.2. Schematic representation of tropocollagen particles in the formation of three different structural types of collagen. TC, tropocollagen; SLS, segment long spacing; FLS, fibrous long spacing; ATP, adenosine triphosphate; GP, glycoprotein. (From Gross, *J. Biophys. Biochem. Cytol.*[11])

### 1.2.3. *Water*

The water content of bone varies with the species of the animal, with age, with the nutritional state of the individual, and with the nature of the bone tissue under examination.[13] Perhaps the most important concept concerning

the water of bone is that, while the organic matter of normal compact bone remains relatively constant in relation to volume, calcification of bone occurs by the replacement of water by crystals of bone mineral.[14] The space avail-

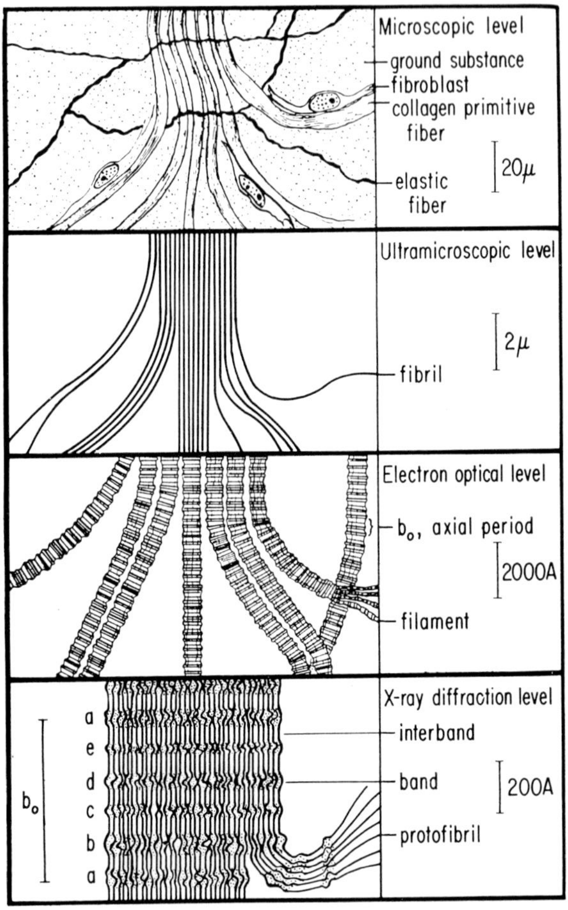


FIG. 2.3. Diagrams of collagenous elements shown at levels of increasing magnification. (From Bear, *Advan. Protein Chem.*[12])

able for mineral is the greater part of that occupied by water; calcification and crystal growth proceed until there is no space left for further expansion. As crystals form and grow in a fixed volume by displacement of water, the spaces between the crystals become smaller and smaller until eventually a state of maximal mineralization is achieved.

## 2. Formation and Organization of Bone

Bone is closely related to cartilage, another specialized form of connective tissue. Most of the embryonic skeleton is laid down first as models of hyaline cartilage, the cells of which hypertrophy and undergo changes in their chemical characteristics immediately prior to their replacement by bone.[15] When bone is formed in such a manner as to replace cartilage either in embryonic or in postfetal life, *endochondral*, or *intracartilaginous ossification* is said to take

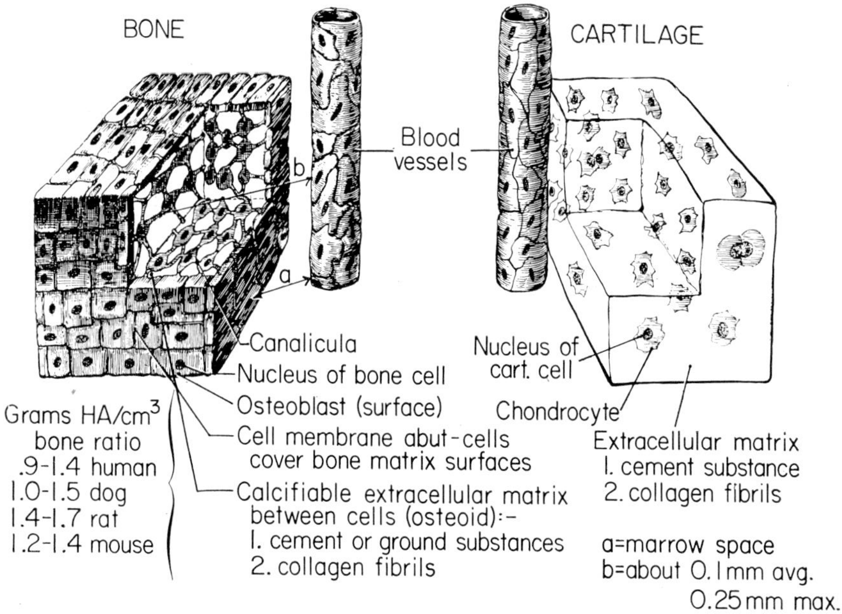


FIG. 2.4. A diagrammatic comparison between bone and cartilage. HA, hydroxyapatite. (From Robinson, *Clin. Orthopaed.*[14])

place. When the transformation is direct without the presence of cartilage, it is called *intramembranous ossification.*

Bones are organized on two levels: (1) as a tissue, and (2) as organs. The organization of bone as a tissue[17] is displayed in the structure of the organic matrix, and in the relation of this matrix to the osteocytes, with their lacunae and interconnecting canalicules. The fibers of compact bone, instead of being arranged in a random fashion, are oriented to the structure of the haversian systems.

In primary ossification, beginning in embryonic life, the fibers of bone are interwoven, without a definite internal structure. Subsequently, tunnels are hollowed out in the osseous tissue, by resorption from the marrow cavity—

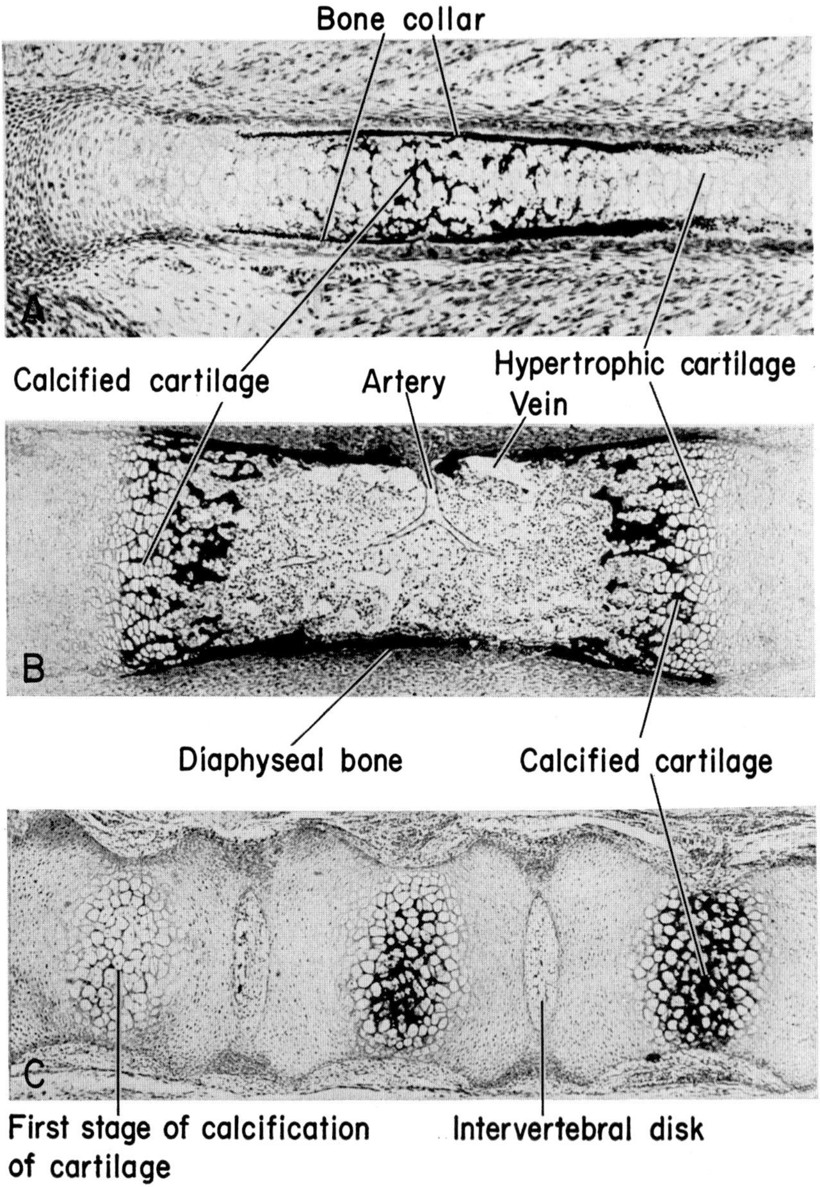


FIG. 2.5. A, longitudinal section through an 18-day-old rat embryo rib. B, section of 4-day-old rat metatarsal. C, section through cartilage models of 20-day rat embryo vertebral bodies showing three stages in the calcification of the cartilage model. Silver nitrate-hematoxylin-eosin stain. (Magnification × 107, 58, and 53.) (From Bloom and Bloom, *Anat. Record*.[15])

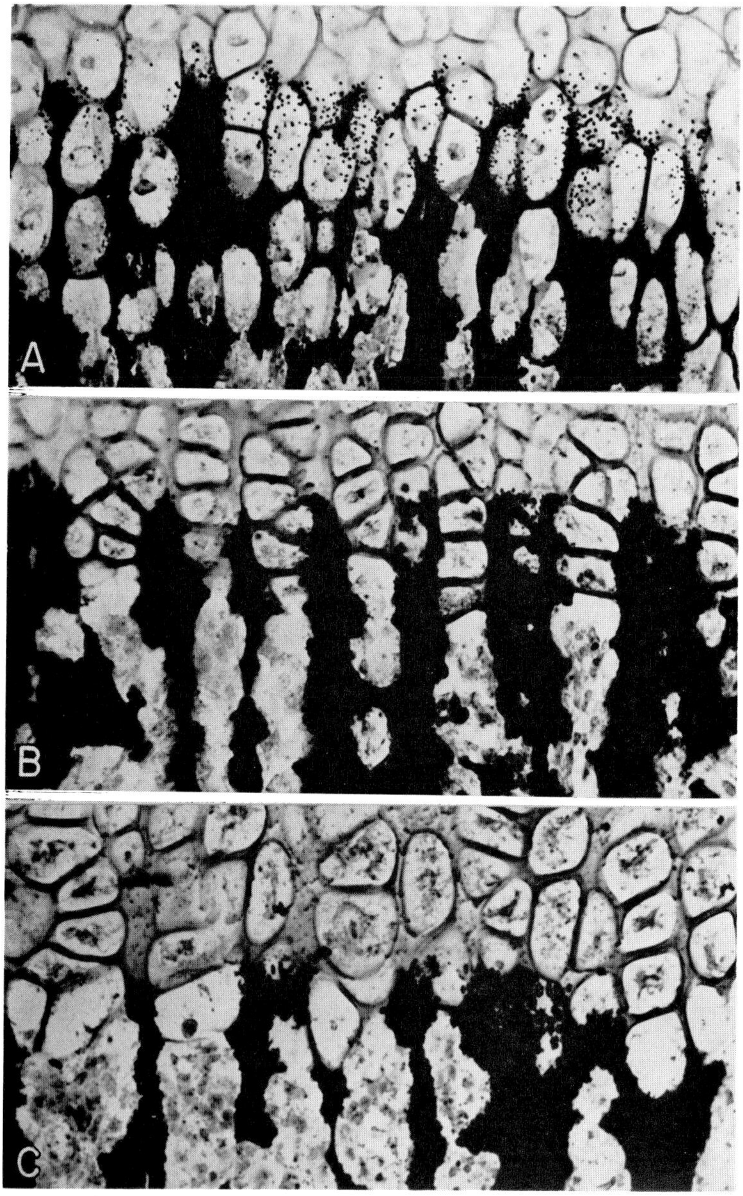


FIG. 2.6. Sections through zones of hypertrophic cartilage and provisional calcification in epiphyseal cartilages. A, proximal end of tibia of a normal 28-day-old rat. B, distal end of radius of normal 34-day-old puppy. C, costochondral junction, same puppy. Silver nitrate-hematoxylin-eosin stain. (Magnification × 262.) (From McLean and Bloom, *Anat. Record*.[16])

these tunnels are known as *primitive haversian systems.* After the age of 1 year in man all new bone formed in the diaphyses of the long bones is laid down in layers, or lamellae.

The unit of structure of compact bone is the haversian system, or osteon. This is, when fully formed, an irregularly cylindrical and branching structure, with thick walls and a narrow lumen—the haversian canal. The canal carries one or more blood vessels, mainly capillaries and venules. The cylindrical osteons are usually oriented in the long axes of the bones. The walls have a definite lamellar structure, the fibrils of each lamella running spirally to the axis of the canal; the direction of the fibrils changes from layer to layer of the successive lamellae. The haversian system, in addition to being arranged around a central canal, includes large numbers of lacunae which house the osteocytes and interconnect with one another and with the lumen of the canal by means of branched canalicules.

The primitive or primary haversian system, or osteon, with its haversian canal represents the initial state of *compact bone.* In some animals such as the mouse and rat, organization of bone does not progress beyond this stage. In larger animals, *secondary osteons* are produced by formation of the cylindrical *absorption cavities* of Tomes and De Morgan[18] (resorption cavities), by deposition of concentric layers of lamellae of bone, again leaving, in each case, a haversian canal. Successive generations of haversian systems are then formed throughout the life of the animal. They cut across the lines of the earlier osteons, leaving irregularly shaped areas of *interstitial lamellae.* Additional *circumferential lamellae* are formed under the periosteum and endosteum. Compact bone thus consists of lamellar bone, arranged in a variety of ways: in relatively new and regularly shaped osteons; in portions of older osteons, remaining as interstitial lamellae; and in circumferential layers near the surfaces. All of these lamellae are similar in structure and alike in function; their differences are incidental to the sequences occurring in their formation.

The resorption cavity or tunnel, as seen in sections, contains blood vessels and connective tissue cells. While resorption is in progress, the cavity is lined with osteoclasts. When resorption ceases and rebuilding begins, the osteoclasts are replaced by osteoblasts. The tunnel is then filled in, from its walls toward the center, by apposition of bone in successive concentric layers or lamellae, a process which incorporates the osteocytes, their lacunae, and the canalicules. Wherever the central blood vessel branches, a lateral branch of the osteon is formed to surround it; some of these branches become the canals of Volkmann and communicate with the marrow cavity or the periosteal surface of the bone. The formation of new layers continues until the canal reaches its final diameter, usually approximately 20 $\mu$.

Differences in density of individual osteons are clearly demonstrated by microradiography[20]; the younger osteons are less mineralized and, consequently, less dense than the older ones. The new and less dense osteons account for the more prominent portion of the uptake of radioactive calcium, strontium, and phosphorus, recognizable in autoradiograms as *hotspots*.

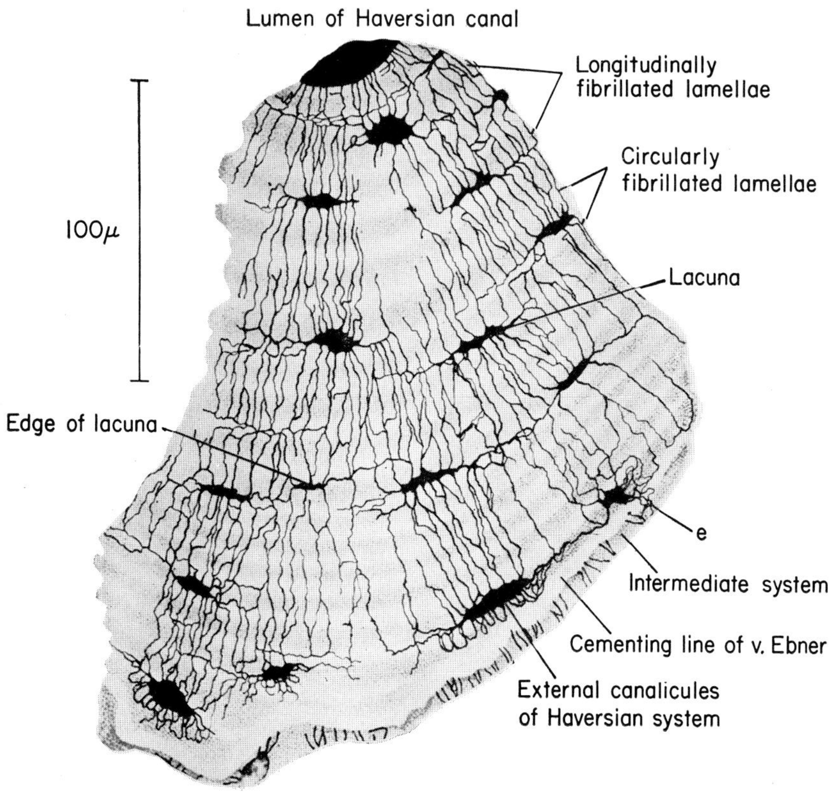


FIG. 2.7. Sector of a cross section of a haversian system of a macerated human hip bone. e, Connection of canalicules of the haversian system with those of an intermediate system. (From Bloom and Fawcett, *A Textbook of Histology*.[19])

## 3. The Mineral of Bone

The hardness of bone—its outstanding characteristic—results from its content of a mineral, deposited in the organic matrix. The mineral, commonly called the bone salt, and the organic matrix, together with the associated water, make up the interstitial substance of bone. Many of the contributions to understanding of the bone mineral, and of its role in metabolism, have been made

by application of tracer methods, and will be examined in detail in subsequent chapters.

X-ray diffraction, together with chemical analysis, has brought about general agreement that the basic structure of bone mineral is that of hydroxyapatite, whose formula may be approximated as $Ca_{10}(PO_4)_6(OH)_2$. The uncertainties in the chemical composition will receive further consideration in Chapter 5.

## 4. Teeth and Supporting Tissues

### 4.1. The Tooth as a Tissue

Just as bone is organized on two levels, a tooth can be considered to have similar levels of organization, i.e., as a tissue and as an organ. Radioisotopes and other modern experimental tools have added little to the existing knowledge of descriptive histology. However, modern techniques have rapidly altered the conventional concepts regarding the function and chemical stability of dental hard tissues.[21] It will be apparent that while bones and teeth differ as organs, as tissues they have many similarities (Table 2.1). While the dental hard tissues, with the exception of cementum, are not normally subject to the reconstruction and remodeling of bone, the presence of fewer metabolic compartments in teeth affords an opportunity to elucidate certain of the aspects of hard tissue metabolism which are shared with bone, but not readily singled out in studies on bone alone.

### 4.2. Histogenesis of the Teeth

The enamel of teeth is a product of the ectodermal epithelium; all the other parts are derivatives of the connective tissue. During the formation of the teeth, the dentin, pulp, and periodontal membrane (all of mesenchymal origin) become fused with the enamel (of ectodermal origin) to form single structures.

In human embryos of the fifth week, the ectodermal epithelium lining the oral cavity presents a thickening along the edge of the future upper and lower jaws. The thickening consists of two solid epithelial ridges which extend into the subjacent mesenchyme. Of these, the labial ridge (nearer the tongue) produces the teeth and is called the *dental lamina.* The edge of the dental lamina produces budlike thickenings—the primordia of the teeth—the *tooth germs;* these become the *enamel organs.* The tissue enclosed by the enamel organ is a denser mesenchyme, constituting the *dental papilla;* the convex surface of the papilla is capped by the bell-shaped enamel organ. The internal cells of the enamel organ differentiate, and, as a result of an accumulation of gelatinous intercellular substance, they constitute a *stellate*

TABLE 2.1
COMPARISON OF MINERALIZED TISSUES[a]

| | Bone | Cementum | Dentin | Enamel |
|---|---|---|---|---|
| *Development:* | | | | |
| Embryonic origin | Mesenchymal | Mesenchymal | Mesenchymal | Epithelial |
| Ossification | Endochondral, intramembranous | Intramembranous | Intramembranous | Secretory |
| Uncalcified precursor | Osteoid | Cementoid | Predentin | Pre-enamel |
| Appositional pattern | Lamellar | Lamellar | Incremental lines of Ebner | Incremental lines of Retzius |
| Structural type | Spongy, compact | Cellular, acellular | Tubular | Prismatic |
| *Environment:* | | | | |
| Internal vascular canals | Haversian, Volkmann's | None | None | None |
| Internal cellular spaces | Lacunae, Canaliculi | Lacunae, Canaliculi | Dentinal tubules | None |
| Internal cellular elements | Osteocytes | Cementocytes | Odontoblastic processes | None |
| Adjacent cellular elements | Fibroblasts, osteoblasts, osteoclasts | Fibroblasts, cementoblasts | Fibroblasts, odontoblasts | None |
| Adjacent tissue fluid environment | Connective tissue fluid | Connective tissue fluid | Connective tissue fluid | Saliva |
| *Main Components:* | | | | |
| Inorganic crystals | Apatite | Apatite | Apatite | Apatite |
| Ground substance | Polysaccharide | Polysaccharide | Polysaccharide | Polysaccharide |
| Fibrous framework | Collagen | Collagen | Collagen | Keratin |

[a] From Sognnaes, *Ann. N.Y. Acad. Sci.*[22]

*reticulum.* Toward the oral cavity, the stellate reticulum is bounded by the outer enamel epithelium, a single layer of cuboidal cells; toward the dental papilla, it is bounded by the inner enamel epithelium, the cells of which elongate and become the enamel-producing cells, the *ameloblasts.*

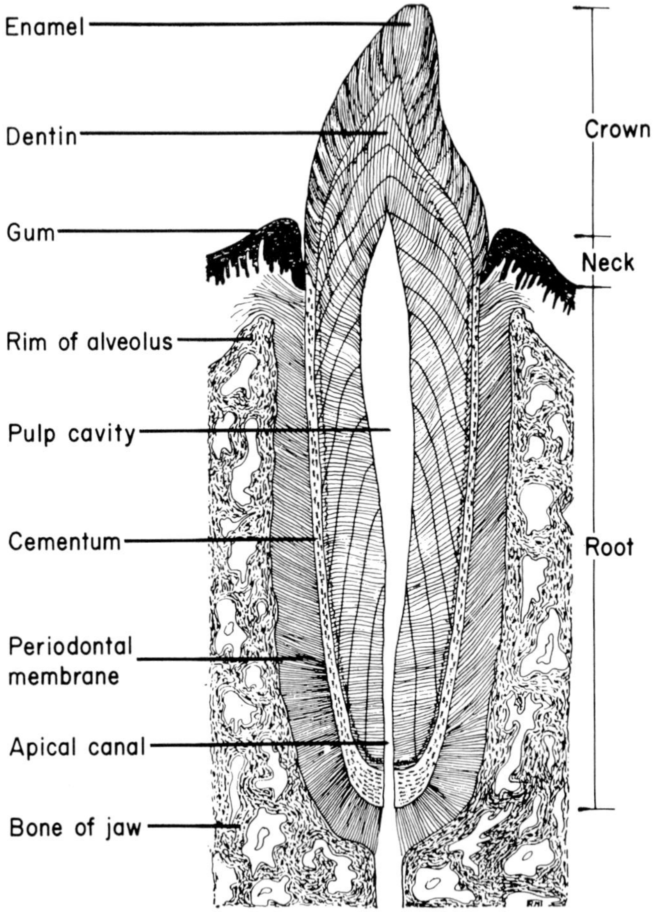


FIG. 2.8. Section of a tooth in position in the jaw. (From Sognnaes, *Histology.*[23])

When the formation of the hard tooth substance begins (in human embryos of about 20 weeks), the mesenchyme of the papilla contains numerous blood vessels and a few reticular fibers between its cells. The cells adjacent to the layer of ameloblasts become transformed into *odontoblasts.* The *dentin* first appears in a thick limiting line between ameloblasts and odontoblasts. The odontoblasts are concerned in the nourishment and possibly in the deposition

of calcium in the dentin; whether or not they actually form the dentin is uncertain. Just before the dentin is formed, the odontoblasts develop large amounts of glycoprotein that is possibly related to the ground substance. The layer of dentin gradually grows thicker and is transformed into a solid cap by the apposition of new layers on its concave surface. In dentin formation, calcification follows closely on deposition of the fibrillar substance. During the whole process, there is a thin layer of uncalcified dentin near the odontoblasts—the *predentin.*

Almost immediately after the appearance of the first calcified dentin on the convexity of the papilla, the ameloblasts begin the elaboration of enamel; it is deposited layer by layer on the surface of the calcifying dentin. The formation of enamel begins at the top of the crown of each tooth and spreads downward over its sides. At a distance from the ameloblasts, the enamel matrix becomes calcified and consists of rods known as *enamel prisms,* approximately 5 μ thick; these are enveloped by the *prism sheaths* and bound together by a calcified *interprismatic substance.* The enamel increases in thickness by elongation of the prisms, and the ameloblasts recede in the direction of the enamel organ. The prisms become slightly broader as they develop radially; by the time the tooth is ready to erupt, the enamel organ has ceased to function and is much reduced in structure. The organ then degenerates and becomes applied to the surface of the erupting tooth, forming the dental cuticle.[19,23]

The erupted tooth, as shown in Fig. 2.8, consists of: a *crown,* the portion which projects above the *gingiva* into the oral cavity; the *neck,* the portion which is covered by the gingiva; and the *root,* the part inserted into the socket or *alveolus* in the bone of the jaw. The hard portion of the tooth consists of three calcified substances: *enamel, dentin,* and *cementum.* The cementum together with the *periodontal membrane* and gingiva serves as the supporting tissue of the tooth.

## 4.3. Dental Tissues

Like bone, the dental tissues consist of a densely mineralized organic matrix. Unlike bone, however, the dentin and enamel are acellular and avascular. Embryologically, the dentin, cementum, pulp, and periodontal membrane of the teeth are of mesenchymal origin like bone and other connective tissues (see Table 2.1).

### *4.3.1. Enamel*

This cuticular formation of epithelial origin is the hardest substance found in the body. It is a highly specialized ectodermal structure. When fully de-

veloped, enamel consists almost entirely of calcium salts in the form of large apatite crystals, while only 3% of it is organic substance containing a glycoprotein.

In fully calcified enamel, the matrix is calcified and contains from 96 to 98% inorganic matter. Enamel contains no cells or other protoplasmic structures. Striations in the enamel can be demonstrated as regularly spaced thickening of the prism sheaths along areas of enamel formation. These striations indicate the shape of the entire enamel at successive stages in development and are called *contour lines* or *lines of Retzius*. The contour lines may be interpreted as growth or incremental lines. The physiological change in metabolism at birth stamps an imprint on the teeth; this is the *neonatal line*.

Ultrathin sections examined under the electron microscope reveal a submicroscopic organic framework within and between the enamel prisms. This delicate network envelops the individual inorganic crystals.

Radioisotopes have been especially useful in the study of amelogenesis; as the crystals grow, they offer a progressively increasing surface area for the deposition of more calcium and phosphate ions. This will be given further treatment in Chapter 5.

### *4.3.2. Dentin*

This is a fibrillar, calcified collagenous matrix traversed by odontoblastic processes composed of a proteinlike elastin.[24] Dentin is harder than compact bone, although it resembles bone in its structure, chemical nature, and development.[19] Minerals enter the dentin of young animals along the predentinodentinal junction; new layers of minerals are continuously added to dentin by accretion, and this continues in adult animals at a slower rate.[25] The cells of dentin remain on the surface of the intercellular substance and send their processes into it. As in bone, the substance of macerated dentin consists of 28% organic and 72% inorganic fractions.[19]

Dentin continues to be formed slowly throughout life and the pulp cavity becomes reduced in size with age. Injury causes an increased activity and the deposit of new or *secondary dentin*.

### *4.3.3. Pulp*

The pulp originates from the dental papilla which is composed of condensed mesenchyme. The young papilla is very cellular; the cells are round or polyhedral and moderately large with pale, almost unstained cytoplasm and large nuclei. In man these round cells are found only in the pulp of fetal teeth; they appear to have no counterpart in other connective tissues unless it be cartilage cells. As the pulp matures, the round cells of the dental papilla become spindle-

shaped and their metachromasia diminishes whereas an intense phosphatase reaction persists.

The adult pulp consists of a syncytium of spindle-shaped mesenchymal cells (with the characteristics of fibroblasts) together with a peripheral layer of odontoblasts. The intercellular spaces are filled with a viscous fluid and a fine network of reticular fibers. The pulp is free from elastic and collagenous fibers and is very vascular.

## 4.4. Supporting Tissues

The cells of the inner and outer enamel epithelium together form the *epithelial sheath* of the root. This sheath determines the extent and shape of the downward growth of the root. Each tooth germ, consisting of its enamel organ and papilla, is completely surrounded by mesenchyme, forming the *dental sac*. After the dentin of the root is formed, the ectodermal layers degenerate and leave the outer surface of the dentin exposed to the surrounding mesenchyme. The mesenchymal cells become similar to osteoblasts; these cells, the *cementoblasts*, form a modified appositional bone on the dentinal surface, the cementum.

### *4.4.1. Cementum*

This is the coarsely fibrillated, interstitial bone substance. A thin layer of cementum is deposited before the tooth erupts. This layer is noncellular and is called the *primary cementum*. After eruption, a new set of cells forms cementum against the external root surface, and, at this time when little or no movement of the tooth disturbs the cementoblasts, the *secondary cementum* becomes much thicker and contains several layers of cells, the *cementocytes*. The lamellae of the cementum, which are seldom well marked, are concentrically placed around the root. Haversian canals are generally absent but may occur in old teeth in the outer layers of cementum near the apex of the root. In addition to supplying cementoblasts to form the secondary cementum, the mesenchyme of the dental sac also produces thick collagenous fibers which become embedded in the cementum matrix. These are *Sharpey's fibers*, comparable to those in bone.

### *4.4.2. Periodontal Membrane*

This structure also serves as the periosteum to the alveolar bone and furnishes a firm connection between the root and the bone. The membrane differs from the skeletal periosteum by the absence of elastic fibers. In addition to the cementoblasts and Sharpey's fibers, the periodontal membrane contains the usual constituents of dense collagenous connective tissue.

### *4.4.3. Gingiva*

This tissue is firmly connected with the periosteum at the crest of the alveolar bone. Histologically, gingival tissue resembles the epidermis of the skin in many respects. The ground substance of the gum and periodontal membrane is metachromatic; the stroma of the gingival tissue varies in this regard, but in places it is deeply metachromatic and contains exceptionally large numbers of mast cells.

## REFERENCES

1. A. Dorfman, The Biochemistry of Connective Tissue, *J. Chronic Disease*, **10**: 403-417 (1959).
2. K. Meyer, The Mucopolysaccharides of Bone, in *Ciba Foundation Symposium on Bone Structure and Metabolism*, pp. 65-74, Little, Brown & Co., Boston, 1956.
3. I. Gersh, Ground Substance and the Plasticity of Connective Tissues, *Harvey Lectures, Ser. XLV*, 211-241 (1952).
4. M. Heller-Steinberg, Ground Substance, Bone Salts, and Cellular Activity in Bone Formation and Destruction, *Am. J. Anat.*, **89**: 347-379 (1951).
5. F. C. McLean and M. R. Urist, Newer Knowledge of Post-fetal Osteogenesis, *Bull. Hosp. Joint Diseases*, **12**: 431-442 (1951).
6. F. C. McLean and W. Bloom, Calcification and Ossification. Mobilization of Bone Salt by Parathyroid Extract, *Arch. Pathol.*, **32**: 315-333 (1941).
7. W. Bloom, M. A. Bloom, and F. C. McLean, Calcification and Ossification. Medullary Bone Changes in the Reproductive Cycle of Female Pigeons, *Anat. Record*, **81**: 443-475 (1941).
8. M. R. Urist, A. M. Budy, and F. C. McLean, Endosteal-bone Formation in Estrogen-treated Mice, *J. Bone Joint Surg.*, **32-A**: 143-162 (1950).
9. F. Wassermann, *Electron Microscopic Examination of the Wall of the Lacunae and Canaliculi in Bone*, USAEC Report ANL-6535, pp. 129-138, Argonne National Laboratory, April 1962.
10. M. R. Urist and F. C. McLean, Accumulation of Mast Cells in Endosteum of Bones of Calcium-deficient Rats, *A.M.A. Arch. Pathol.*, **63**: 239-251 (1957).
11. J. Gross, The Behavior of Collagen Units as a Model in Morphogenesis, *J. Biophys. Biochem. Cytol.*, **2** Suppl.: 261-274 (1956).
12. R. S. Bear, The Structure of Collagen Fibrils, *Advan. Protein Chem.*, **VII**: 69-160 (1952).
13. R. A. Robinson and S. R. Elliott, The Water Content of Bone. I. The Mass of Water, Inorganic Crystals, Organic Matrix and "$CO_2$ space" Components in a Unit Volume of Dog Bone, *J. Bone Joint Surg.*, **39-A**: 167-188 (1957).
14. R. A. Robinson, Crystal-collagen-water Relationships in Bone Matrix, *Clin. Orthopaed.*, **17**: 69-76 (1960).
15. W. Bloom and M. A. Bloom, Calcification and Ossification, Calcification of Developing Bones in Embryonic and Newborn Rats, *Anat. Record*, **78**: 497-523 (1940).
16. F. C. McLean and W. Bloom, Calcification and Ossification. Calcification in Normal Growing Bone, *Anat. Record*, **78**: 333-359 (1940).
17. P. Lacroix, *The Organization of Bones*, Blakiston Co., Philadelphia, 1951.
18. J. Tomes and C. De Morgan, Observations on the Structure and Development of Bone, *Phil. Trans. Roy. Soc. London*, **143**: 109-139 (1853).

19. W. Bloom and D. W. Fawcett, *A Textbook of Histology*, 8th ed., pp. 411-426, W. B. Saunders Co., Philadelphia, 1962.
20. R. E. Rowland, Radiography: Micro-; Quantitative, of Bone, in *Medical Physics*, Vol. III, pp. 525-528, The Year Book Publishers, Inc., Chicago, 1960.
21. R. F. Sognnaes, Radioisotope Interactions with Dental Hard Tissues, in *Radioisotopes and Bone*, pp. 355-370, Blackwell Scientific Publications Ltd., Oxford, 1962.
22. R. F. Sognnaes, Microstructure and Histochemical Characteristics of the Mineralized Tissues, *Ann. N. Y. Acad. Sci.*, **60**: 545-574 (1955).
23. R. F. Sognnaes, Oral Cavity, in *Histology*, pp. 458-488, Blakiston Co., New York, 1954.
24. M. V. Stack, The Chemical Nature of the Organic Matrix of Bone, Dentin, and Enamel, *Ann. N. Y. Acad. Sci.*, **60**: 585-595 (1955).
25. C. P. Leblond, L. F. Bélanger, and R. C. Greulich, Formation of Bones and Teeth as Visualized by Radioautography, *Ann. N. Y. Acad. Sci.*, **60**: 631-659 (1955).

## Bibliography

W. D. Armstrong, Biochemical and Nutritional Studies in Relation to the Teeth, *Ann. Rev. Biochem.*, **11**: 441-464 (1942).

W. Bloom and D. W. Fawcett, *A Textbook of Histology*, 8th ed., W. B. Saunders Co., Philadelphia, 1962.

F. C. McLean and M. R. Urist, *Bone: An Introduction to the Physiology of Skeletal Tissue*, 2nd ed., The University of Chicago Press, Chicago, 1961.

R. A. Robinson and D. A. Cameron, The Organic Matrix of Bone and Epiphyseal Cartilage, *Clin. Orthopaed.*, **9**: 16-29 (1957).

G. B. Wislocki, H. Bunting, and E. W. Dempsey, Metachromasia in Mammalian Tissues and Its Relationship to Mucopolysaccharides, *Am. J. Anat.*, **81**: 1-37 (1947).

# Chapter 3

# *Bone-Seeking Elements*

## 1. Introductory Statement

In its broadest sense, a bone seeker is any substance localizing in the skeleton; this would include vital dyes used for staining of matrix which persist after decalcification and substances such as alizarin and tetracycline which combine with bone mineral in new calcification sites.[1] In this chapter, bone seekers will be limited to those elements and their radionuclides which are deposited and retained in the skeleton.

For orientation among the elements, there is presented a periodic table (Fig. 3.1) which includes the symbols and mass ($A$) numbers of all known elements; the shaded areas indicate those deposited in the skeleton.[1-5] More detailed information concerning the radioisotopes of these elements, with some indication of their usefulness as tracers for the study of bone and cartilage, is given below.

## 2. Periodic Table of Elements

Of first interest in this table are the divalent cations of the elements in Group IIA (the alkaline earths) which include Ca, Sr, and Ra, all of prime importance in connection with bone, together with Be and Ba, of some interest but of lesser importance. Another point of interest is the quadrangle (Groups IA and IIA) which includes Na, K, Mg, and Ca; these are the four most abundant and physiologically important elements which appear as cations in the animal organism, and there are many interrelations between them. It is worthy of special note that, while Ca is the basis of the crystal structure of bone mineral, Na plays only a secondary role in this connection, the place of Mg in bone is uncertain, and K shows no special affinity for bone.

In Groups V and VI, P and S will receive special attention but for different reasons. Phosphorus, in the form of phosphate, shares with Ca an important place in the bone mineral and plays a significant role in many physiological activities; the economy of phosphorus in the animal organism is second to no other element in interest and importance. Sulfur, as sulfate, is incorporated in the sulfated mucopolysaccharides of the ground substance of bones and teeth.[6] The radionuclide $S^{35}$ has proved useful as a tracer element.

| Periods | I (A / B) | II (A / B) | III (A / B) | IV (A / B) | V (A / B) | VI (A / B) | VII (A / B) | VIII | O |
|---|---|---|---|---|---|---|---|---|---|
| I | H 1 | | | | | | | | He 2 |
| II | Li 3 | Be 4 | 5 B | 6 C | 7 N | 8 O | 9 F | | Ne 10 |
| III | Na 11 | Mg 12 | 13 Al | 14 Si | 15 P | 16 S | 17 Cl | | Ar 18 |
| IV | K 19 / 29 Cu | Ca 20 / 30 Zn | Sc 21 / 31 Ga | Ti 22 / 32 Ge | V 23 / 33 As | Cr 24 / 34 Se | Mn 25 / 35 Br | Fe 26 / Co 27 / Ni 28 | Kr 36 |
| V | Rb 37 / 47 Ag | Sr 38 / 48 Cd | Y 39 / 49 In | Zr 40 / 50 Sn | Nb 41 / 51 Sb | Mo 42 / 52 Te | Tc 43 / 53 I | Ru 44 / Rh 45 / Pd 46 | Xe 54 |
| VI | Cs 55 / 79 Au | Ba 56 / 80 Hg | La 57--- / 81 Tl | Hf 72 / 82 Pb | Ta 73 / 83 Bi | W 74 / 84 Po | Re 75 / 85 At | Os 76 / Ir 77 / Pt 78 | Rn 86 |
| VII | Fr 87 | Ra 88 | Ac 89--- | | | | | | |

| | | | | | | | | | | | | | | |
|---|---|---|---|---|---|---|---|---|---|---|---|---|---|---|
| Lanthanides | Ce 58 | Pr 59 | Nd 60 | Pm 61 | Sm 62 | Eu 63 | Gd 64 | Tb 65 | Dy 66 | Ho 67 | Er 68 | Tm 69 | Yb 70 | Lu 71 |
| Actinides | Th 90 | Pa 91 | U 92 | Np 93 | Pu 94 | Am 95 | Cm 96 | Bk 97 | Cf 98 | Es 99 | Fm 100 | Md 101 | No 102 | |

FIG. 3.1. Periodic chart of elements. Those useful in the study of bones and teeth are indicated by shaded areas.

Other elements, and especially their radioactive fission products, of interest are gallium, yttrium, zirconium, and niobium. Certain members of the rare earths, including lanthanum and the lanthanide series, qualify as bone seekers and are still under study with respect to bone. Of very great current interest are actinium and the actinide series, especially since the actinide series includes the transuranic elements, all of which are man-made. Plutonium—a member of this group—has assumed special importance with respect to nuclear fission (see Chapter 11).

All of the heavy elements beyond element 82 (Pb) in the periodic table have radioactive isotopes which emit alpha particles, although some of them exist also in stable forms. Their importance with respect to bone depends upon the possibilities of exposure to their effects and the ease with which they gain access to the skeletal system. Because some of the heavy elements beyond plutonium in the actinide series are being made in increasing amounts in the laboratory, they constitute a potential hazard in relation to bone.

Apart from the groups of elements mentioned, there are approximately 22 elements that are constantly present in the animal organism in very low but variable concentrations; these are commonly designated as *trace elements*. Of these, Mn, Zn, Cu, and I are essential for normal skeletal development; molybdenum is related to copper metabolism and shows effects on the skeleton when present in the diet in excess.[7] The metabolic role of these essential trace elements will be discussed in Chapter 6. Spectrographic analyses of human bone and cartilage have revealed the presence of the following elements: B, Al, Si, Ti, Cr, Mn, Fe, Co, Ni, Cu, Sr, Mo, Sn, Pb, Bi, Cd, and Ag.[8] The remaining elements as indicated by the shaded areas in Fig. 3.1 have been demonstrated in bone following administration of their radionuclides.[1-4]

The radioisotopes of elements of special interest in relation to the skeleton are summarized in Table 3.1. The isotopes included in this table have assumed importance either because they are useful as tracers in the study of bone or because they constitute a hazard to man. The table includes the half-lives of the elements listed, the types of radiation emitted and their energies, the types of decay and transition products,[10] and some information concerning the interest in relation to bone.

## 3. Transmutation and Translocation

Consideration has been given in Chapter 1 to the transmutation of elements which may occur by beta decay ($\beta^-$ or $\beta^+$), by alpha decay ($\alpha$), or by electron capture (EC). In each instance, the new element formed

TABLE 3.1
RADIOISOTOPES OF SPECIAL INTEREST IN RELATION TO THE SKELETON[a]

| Z No. | Element | Isotope | Half-life[b] | Radiation type and energy, Mev[c] | Decay, type, and transition product[b,d] | Interest for bone |
|---|---|---|---|---|---|---|
| 1 | Hydrogen (tritium) | $H^3$ | 12.3 y | β− 0.018 | β− to stable helium-3 | $H^3$-amino acid to label cell constituents |
| 4 | Beryllium | $Be^7$ | 53.6 d | γ 0.478 | EC to stable lithium-7 | Fixed on crystal surfaces |
| 6 | Carbon | $C^{14}$ | 5568 y | β− 0.155 | β− to stable nitrogen-14 | Fixed as $CO_3^{2-}$ |
| 9 | Fluorine | $F^{18}$ | 109.7 m | β+ 0.649 | β+ and EC to stable oxygen-18 | Substitutes for $OH^-$ |
| 11 | Sodium | $Na^{22}$ | 2.6 y | β+ 0.540<br>γ 1.277 | β+ and EC to stable neon-22 | Deposits on crystal surfaces |
| 12 | Magnesium | $Mg^{28}$ | 21.4 h | β− 0.459<br>γ 0.032–1.35 | β− to 2.3-m radioaluminum-28 | Fixed on crystal surfaces |
| 15 | Phosphorus | $P^{32}$ | 14.3 d | β− 1.707 | β− to stable sulfur-32 | Exchanges for $P^{31}$ as $PO_4^{3-}$ |
| 16 | Sulfur | $S^{35}$ | 87.1 d | β− 0.167 | β− to stable chlorine-35 | As $SO_4^{2-}$ in ground substance |
| 20 | Calcium | $Ca^{45}$ | 164 d | β− 0.254 | β− to stable scandium-45 | Exchanges for $Ca^{40}$; pure beta emitter |
| | Calcium | $Ca^{47}$ | 4.7 d | β− 0.69, 2.0<br>γ 1.3, 0.83, 0.46 | β− to 3.4-d radioscandium-47 | Exchanges for $Ca^{40}$; gamma emitter |
| 25 | Manganese | $Mn^{54}$ | 300 d | γ 0.940 | EC to stable chromium-54 | Associated with bone mineral; small amount bound to matrix |
| 30 | Zinc | $Zn^{65}$ | 245 d | β+ 0.324<br>γ 1.119 | β+ and EC to stable copper-65 | Deposits in preosseous tissue in formation of new osteons |
| 31 | Gallium | $Ga^{72}$ | 14.2 h | β− 0.56–3.17<br>γ 0.631–2.50 | β− to stable germanium-72 | Deposits in regions of osteogenic activity |
| 38 | Strontium | $Sr^{85}$ | 64 d | γ 0.513 | EC to stable rubidium-85 | Exchanges for $Ca^{40}$; gamma emitter |
| | Strontium | $Sr^{89}$ | 51 d | β− 1.5 | β− to stable yttrium-89 | Exchanges for $Ca^{40}$; beta emitter |
| | Strontium | $Sr^{90}$ | 27.7 y | β− 0.54 | β− to radioyttrium-90 | Fallout hazard |

TABLE 3.1 (*Continued*)

| Z No. | Element | Isotope | Half-life[b] | Radiation type and energy, Mev[c] | Decay, type, and transition product[b,d] | Interest for bone |
|---|---|---|---|---|---|---|
| 39 | Yttrium | $Y^{90}$ | 64 h | $\beta^-$ 2.26 | $\beta^-$ to stable zirconium-90 | Deposits on resorbing and quiescent surfaces |
| | Yttrium | $Y^{91}$ | 58.0 d | $\beta^-$ 1.55 | $\beta^-$ to stable zirconium-91 | Deposits on resorbing and quiescent surfaces |
| 40 | Zirconium | $Zr^{95}$ | 65 d | $\beta^-$ 0.75<br>$\gamma$ 0.708 | $\beta^-$ to 35-d radioniobium-95 | Deposits on resorbing and quiescent surfaces |
| 41 | Niobium | $Nb^{95}$ | 35 d | $\beta^-$ 0.158<br>$\gamma$ 0.758 | $\beta^-$ to stable molybdenum-95 | Deposits on resorbing and quiescent surfaces |
| 56 | Barium | $Ba^{133}$ | 7.2 y | $\gamma$ 0.36 | EC to stable cesium-133 | Exchanges for $Ca^{40}$ on crystal surfaces |
| | Barium | $Ba^{140}$ | 12.8 d | $\beta^-$ 0.48<br>$\gamma$ 0.306–0.540 | $\beta^-$ to 40-h radiolanthanum-140 | Exchanges for $Ca^{40}$ on crystal surfaces |
| 57 | Lanthanum | $La^{140}$ | 40 h | $\beta^-$ 1.3–2.26<br>$\gamma$ 0.093–2.9 | $\beta^-$ to stable cerium-140 | Deposits on resorbing and quiescent surfaces |
| 88 | Radium | $Ra^{226}$ | 1622 y | $\alpha$ 4.777<br>$\gamma$ 0.186 | $\alpha$ to radioradon-222 | Exchanges for $Ca^{40}$; induces neoplasms |
| 89 | Actinium | $Ac^{227}$ | 21.6 y | $\alpha$ 4.94<br>$\beta^-$ 0.046 | 1.2% $\alpha$ to radiofrancium-223<br>99.0% $\beta^-$ to radiothorium-227 | Deposits on resorbing and quiescent surfaces |
| 90 | Thorium | $Th^{234}$ | 24.1 d | $\beta^-$ 0.192, 0.104<br>$\gamma$ 0.090 | $\beta^-$ to 6.7-h radio-protactinium-234 | Fixed on crystal surfaces |
| 92 | Uranium | $U^{233}$ | 162,000 y | $\alpha$ 4.816 | $\alpha$ to radiothorium-229 | Fixed on uranyl ion, $UO_2^{2+}$ |
| 94 | Plutonium | $Pu^{239}$ | 24,400 y | $\alpha$ 4.9–5.15<br>$\gamma$ 0.01–0.05 | $\alpha$ to radiouranium-235 | Deposits on resorbing and quiescent surfaces |

[a] Physical data derived from Strominger *et al.*[10]

[b] Half-lives: m, minutes; h, hours; d, days; y, years.

[c] Radiation type: $\alpha$, alpha particles; $\beta^-$, beta particles (negatrons); $\beta^+$, beta particles (positrons); $\gamma$, gamma radiation; Mev, million electron volts.

[d] Decay: $\beta^-$, $\beta^+$, $\alpha$, EC (electron capture).

has chemical properties differing from those of the parent element. This situation is in contrast to the isotopes of a single element, which for the most part all exhibit the same chemical properties.

Attention must also be given to the effect of transmutation of elements upon their physiological properties and their location within the organism. Some of these transmutations, accompanied by changes in location (translocations), may be of physiological importance; others, because of short half-lives or for other reasons, may not be important to the organism but may be used to illustrate the principles involved.

An example is the transmutation of radiocesium ($Cs^{137}$) by $\beta^-$ decay—$Cs^{137}$ has a half-life of 26.6 years—to stable barium ($Ba^{137}$). Most of the transitions produce a metastable excited state of $Ba^{137}$, which has a half-life of 2.6 minutes, before stable $Ba^{137}$ is reached. From the physiological point of view, it is of importance that cesium behaves in a fashion similar to potassium in the organism, while barium, a member of the alkaline earth series, behaves like calcium. Transmutation from cesium to barium therefore suggests that an element which is distributed intracellularly becomes one that is a bone seeker and which acts like calcium in the bone mineral; its large ion diameter may result in its being forced to the surface of the crystals of the bone salt.

Since $Cs^{137}$ is subject to $\beta^-$ decay and $Ba^{137}$ is stable, this transmutation is equivalent to a continuous, very slow infusion of stable barium, diminishing in rate with the loss by decay and by excretion of the parent $Cs^{137}$. Metastable $Ba^{137}$ emits gamma rays only and can be safely ignored. Cesium-137, when ingested, is distributed largely in muscle; in this respect, it follows potassium. The rate of turnover and of excretion of cesium-137 is relatively rapid, and, although it emits both beta particles and gamma rays, its cumulative radiation effect from fallout is small. Because of its rapid excretion, the amount that could be transformed into stable barium-137 and deposited in the bones is negligible from the standpoint of any possible barium toxicity.

Another transmutation of interest is that of calcium to scandium, i.e., $Ca^{47}$ to $Sc^{47}$ (Fig. 3.2). Calcium-47, a bone-seeking isotope, has a short half-life of 4.7 days. This isotope is being prepared artificially and used in patients for external detection of its distribution in the skeleton by virtue of its gamma-ray emission. Calcium-47 is subject to $\beta^-$ decay to $Sc^{47}$, which again emits beta particles and gamma rays with a short half-life (3.4 days) and ultimately decays to stable titanium-47. The $Ca^{47}$ as administered in clinical studies always includes some $Sc^{47}$, and further transmutation occurs in the organism. Little is known about the biological effects of scandium or of titanium or of the localization of $Sc^{47}$. However, it seems doubtful that any significant amount of either of these elements would

localize elsewhere in the body; it is practically certain that the effects of the infinitesimal amounts of stable titanium remaining in the organism are not harmful.

One of the elements frequently used in physiological experimentation is phosphorus ($P^{32}$), usually as orthophosphate. The physical half-life of $P^{32}$ is 14.2 days, and it is transformed by $\beta^-$ decay to stable sulfur ($S^{32}$) as shown in Fig. 3.2. To the possibility of a deleterious effect on genetic material by $\beta^-$ radiation from $P^{32}$ itself, there is added the possibility of

Mass (A) numbers; ATOMIC (Z) NUMBERS

| Z \ A | 26 | 27 | 28 | 29 | 30 | 31 | 32 | 33 | 34 | 35 | 36 | 37 | 38 |
|---|---|---|---|---|---|---|---|---|---|---|---|---|---|
| 14 | $Si^{26}$ ? | $Si^{27}$ $\beta^+$ | $Si^{28}$ 92.2% | $Si^{29}$ 4.7% | $Si^{30}$ 3.1% | $Si^{31}$ $\beta^-, \gamma$ | $Si^{32}$ $\beta^-$ | | | | | | |
| 15 | | | $P^{28}$ $\beta^+, \gamma$ | $P^{29}$ $\beta^+, \gamma$ | $P^{30}$ $\beta^+$ | $P^{31}$ 100% | $P^{32}$ $\beta^-, \gamma$ | $P^{33}$ $\beta^-$ | $P^{34}$ $\beta^-, \gamma$ | | | | |
| 16 | | | | | | $S^{31}$ $\beta^+$ | $S^{32}$ 95.0% | $S^{33}$ 0.75% | $S^{34}$ 4.215% | $S^{35}$ $\beta^-$ | $S^{36}$ 0.17% | $S^{37}$ $\beta^-, \gamma$ | $S^{38}$ $\beta^-, \gamma$ |

| Z \ A | 37 | 38 | 39 | 40 | 41 | 42 | 43 | 44 | 45 | 46 | 47 | 48 | 49 | 50 |
|---|---|---|---|---|---|---|---|---|---|---|---|---|---|---|
| 19 | $K^{37}$ $\beta^+$ | $K^{38}$ $\beta^+$ | $K^{39}$ 93.1% | $K^{40}$ .01% $\beta^-$,EC | $K^{41}$ 6.9% | $K^{42}$ $\beta^-, \gamma$ | $K^{43}$ $\beta^-, \gamma$ | $K^{44}$ $\beta^-, \gamma$ | $K^{45}$ ? | | | | | |
| 20 | | $Ca^{38}$ $\beta^+$ | $Ca^{39}$ $\beta^+$ | $Ca^{40}$ 96.97% | $Ca^{41}$ EC | $Ca^{42}$ 0.64% | $Ca^{43}$ 0.145% | $Ca^{44}$ 2.06% | $Ca^{45}$ $\beta^-$ | $Ca^{46}$ .0033% | $Ca^{47}$ $\beta^-, \gamma$ | $Ca^{48}$ 0.185% | $Ca^{49}$ $\beta^-, \gamma$ | |
| 21 | | | | $Sc^{40}$ $\beta^+, \gamma$ | $Sc^{41}$ $\beta^+$ | $Sc^{42}$ $\beta^+$ | $Sc^{43}$ $\beta^+, \gamma$ | $Sc^{44}$ $\beta^+, \gamma$ | $Sc^{45}$ 100% | $Sc^{46}$ $\beta^-, \gamma$ | $Sc^{47}$ $\beta^-, \gamma$ | $Sc^{48}$ $\beta^-, \gamma$ | $Sc^{49}$ $\beta^-$ | $Sc^{50}$ $\beta^-, \gamma$ |

FIG. 3.2. Partial chart of nuclides, to show isotopes of phosphorus and of calcium, with their relations to isotopes of neighboring elements. Mass ($A$) numbers are in vertical colunms; atomic ($Z$) numbers are in horizontal rows. Shown are atomic numbers of isotopes, relative abundance in nature (in per cent), and type of radiation emitted. Stable isotopes are underlined. Arrows indicate pathways and direction of spontaneous decay, with nature of decay. Alpha decay does not occur in the elements shown. (From McLean and Urist, *Bone: An Introduction to the Physiology of Skeletal Tissue*[9]; physical data derived from Strominger *et al.*[10])

the breakdown of nucleic acid molecules where $P^{32}$ is incorporated[11] when radiophosphorus changes to radiosulfur.

A further transmutation, perhaps of greater biological significance, is that of strontium to yttrium, i.e., $Sr^{90}$ to $Y^{90}$. Strontium-90 is the radioelement with the greatest hazard from fallout because of its relative abundance among the fission products of nuclear weapons; because of its long half-life (27.7 years); and because of its retention in the bone mineral, where it behaves like calcium. It is a pure beta emitter, and its hazard results from beta-particle emission (see Fig. 3.3). The decay product of $Sr^{90}$ is $Y^{90}$, which also emits beta particles, with very weak gamma rays, and

has a half-life of 64.2 hours. Yttrium-90 decays to stable zirconium ($Zr^{90}$). Both strontium-90 and yttrium-90 are bone-seeking isotopes although they have different biological characteristics. Strontium-90, since it is an alkaline earth, behaves like calcium in the bone mineral; yttrium-90 has a special affinity for the surfaces of mineral exposed in the course of resorption of bone. The physiological interest in the transmutation of $Sr^{90}$ to $Y^{90}$ derives in part from the long half-life of strontium-90 coupled with the short half-life of yttrium-90. This results in an equilibrium; $Y^{90}$ is formed from $Sr^{90}$ at the same rate at which it decays to $Zr^{90}$. As a result, the effective concentration of $Y^{90}$ in bone tissue is kept constant in proportion to $Sr^{90}$. The bone is thus under the influence of both of these isotopes. In considering the hazard of $Sr^{90}$, it is assumed that $Y^{90}$ does not wander very far from the locus of its birth. The dose to bone is equal to the sum of the doses due to strontium-90 and yttrium-90. Strontium-90 supplies only 0.193 Mev per disintegration, while that of yttrium-90 is 0.89 Mev. Strontium-85, with a half-life of 64 days, undergoes transmutation by electron capture to the stable $Rb^{85}$.

Mass (A) numbers / ATOMIC (Z) NUMBERS

| Z | 79 | 80 | 81 | 82 | 83 | 84 | 85 | 86 | 87 | 88 | Z |
|---|---|---|---|---|---|---|---|---|---|---|---|
| 37 | $Rb^{79}$ $\beta^+, \gamma$ | $Rb^{80}$ $\beta^+, \gamma$, EC | $Rb^{81}$ $\beta^+, \gamma$, EC | $Rb^{82}$ $\beta^+$ | $Rb^{83}$ $\gamma$, EC | $Rb^{84}$ $\beta^-, \beta^+$ $\gamma$, EC | $Rb^{85}$ 72.1% | $Rb^{86}$ $\beta^-, \gamma$ | $Rb^{87}$ 27.8% | $Rb^{88}$ $\beta^-, \gamma$ | 37 |
| 38 | | $Sr^{80}$ $\gamma$, [EC] | $Sr^{81}$ $\beta^+$, EC | $Sr^{82}$ EC | $Sr^{83}$ $\beta^+, \gamma$, EC | $Sr^{84}$ 0.5% | $Sr^{85}$ $\gamma$, EC | $Sr^{86}$ 9.87% | $Sr^{87}$ 7.0% | $Sr^{88}$ 82.5% | 38 |
| 39 | | | | $Y^{82}$ | $Y^{83}$ | $Y^{84}$ $\beta^+, \gamma$, EC | $Y^{85}$ $\beta^+$ | $Y^{86}$ $\beta^+, \gamma$ | $Y^{87}$ $\beta^+, \gamma$, EC | $Y^{88}$ $\beta^+, \gamma$, EC | 39 |
| | 79 | 80 | 81 | 82 | 83 | 84 | 85 | 86 | 87 | 88 | |

Mass (A) numbers

| Z | 89 | 90 | 91 | 92 | 93 | 94 | 95 | 96 | 97 | Z |
|---|---|---|---|---|---|---|---|---|---|---|
| 37 | $Rb^{89}$ $\beta^-, \gamma$ | $Rb^{90}$ $\beta^-, \gamma$ | $Rb^{91}$ $\beta^-, \gamma$ | $Rb^{92}$ [$\beta^-$] | $Rb^{93}$ [$\beta^-$] | $Rb^{94}$ [$\beta^-$] | $Rb^{95}$ [$\beta^-$] | | $Rb^{97}$ [$\beta^-$] | 37 |
| 38 | $Sr^{89}$ $\beta^-$ | $Sr^{90}$ $\beta^-$ | $Sr^{91}$ $\beta^-, \gamma$ | $Sr^{92}$ $\beta^-, \gamma$ | $Sr^{93}$ $\beta^-$ | $Sr^{94}$ $\beta^-$ | $Sr^{95}$ [$\beta^-$] | | $Sr^{97}$ [$\beta^-$] | 38 |
| 39 | $Y^{89}$ 100% | $Y^{90}$ $\beta^-$ | $Y^{91}$ $\beta^-, \gamma$ | $Y^{92}$ $\beta^-, \gamma$ | $Y^{93}$ $\beta^-, \gamma$ | $Y^{94}$ $\beta^-, \gamma$ | $Y^{95}$ $\beta^-$ | | $Y^{97}$ [$\beta^-$] | 39 |
| | 89 | 90 | 91 | 92 | 93 | 94 | 95 | 96 | 97 | |

FIG. 3.3. Partial chart of nuclides, to show isotopes of strontium and of yttrium, with their relations to isotopes of neighboring elements. Mass ($A$) numbers are in vertical columns; atomic ($Z$) numbers are in horizontal rows. Shown are atomic numbers of isotopes, relative abundance in nature (in per cent), and type of radiation emitted. Stable isotopes are underlined. Arrows indicate pathways and direction of spontaneous decay, with nature of decay. Alpha decay does not occur in the elements shown. (Physical data derived from Strominger *et al.*[10])

One of the radioisotopes listed as of possible concern from fallout is $Ba^{140}$, as judged by the criteria of the amount released into the environment, the physical half-life, the efficiency of transfer through the food chain to the human diet, the degree of absorption by the body, and the biological half-life. Barium-140 has a physical half-life of 12.8 days and behaves in the body like calcium, being accumulated in the bone mineral. Because of the time required for this pathway, the passage of $Ba^{140}$ through the soil and its short half-life reduce this hazard. Barium-140 passes by means of $\beta^-$ decay to lanthanum ($La^{140}$), also a beta and gamma emitter with a half-life of 40.2 hours. Lanthanum-140 then decays further to stable cerium ($Ce^{140}$). Although lanthanum and cerium are both bone seekers, neither element is absorbed to a significant degree by the gastrointestinal tract. Barium, however, is so absorbed and may therefore serve as a means for introducing lanthanum and cerium into bone. Both lanthanum and cerium have a specific affinity for the matrix, even in the absence of the bone mineral. While their distribution in bone has been well studied,[12] their physiological effects, especially in the quantities that might be expected to derive from $Ba^{140}$, have not been determined.

Another type of transmutation with translocation, of potential biological significance, is the formation of $C^{14}$ from $N^{14}$, bombarded in the atmosphere by neutrons ($n^1$) released by detonation of a nuclear weapon. The reaction is:

$$N^{14} + n^1 \rightleftharpoons C^{14} + H^1$$

The $C^{14}$ is diffused in the atmosphere as $C^{14}O_2$ and may reach man either directly by inhalation or indirectly through plants. $C^{14}O_3{}^{2-}$ has been shown experimentally to be deposited in bone, where about one-half of the labeled carbonate remains for a very long time[13]; it has not been demonstrated in bone as a consequence of fallout, and the hazard to man from deposition of this isotope in bone has not been regarded as significant. A more important hazard is perhaps the inclusion of radiocarbon in the genetic material of the human body. It has been concluded[14] that the transmutation effect of carbon-14 atoms contained in the genetic material of the human body could possibly lead to about the same number of genetic mutations as the radiation effect from radiocarbon. The transmutation from $N^{14}$ to $C^{14}$ takes place in the atmosphere and $C^{14}$ enters the body as such. In the body, $C^{14}$ decays to $N^{14}$, and it is this transmutation that may lead to genetic effects.

### 3.1. Transmutation and Translocation in the Heavy Elements

There exist in nature four series of radioactive elements, all of which originate in long-lived heavy elements and decay by emission of both alpha

and beta particles. In each series, the end product is a stable heavy metal —lead or bismuth. The four series are:

1. The thorium series: $Th^{232}$ to $Pb^{208}$
2. The neptunium series: $Np^{237}$ to $Bi^{209}$
3. The uranium-radium series: $U^{238}$ to $Pb^{206}$
4. The uranium-actinium series: $U^{235}$ to $Pb^{207}$

Of these four series, two have assumed biological importance—the uranium-radium series and the thorium series. Members of both of these series have been introduced into the human organism (either accidentally or by design) where they have caused damage. We are interested here in exploring the hazard to man of the radioisotopes in these two series and, more particularly, the translocations that may occur as a result of the transmutation of one element to another with different chemical properties and different points of attack in the organism.

Unfortunately the relationship of transmutation to translocation among the heavy elements is unnecessarily obscured by the outmoded system of nomenclature which still pervades the literature. If one recalls that all isotopes of each of the elements exhibit comparable chemical and physiological characteristics and if the isotopes are designated by their chemical symbols and mass numbers, then it is not difficult to follow the changes in behavior as an atom of one element decays into that of another. We shall follow here the standard nomenclature for the elements listed, including the alternate symbols where desirable for purposes of identification.

In both the uranium-radium series and the thorium series, the most troublesome complications are introduced by the appearance of isotopes of both radium and of thorium in each series. Thus the uranium-radium series includes $Ra^{226}$ and $Rn^{222}$ as well as $Th^{234}$ and $Th^{230}$, while the thorium series includes $Ra^{228}$, $Ra^{224}$, and $Rn^{220}$ together with $Th^{232}$ and $Th^{228}$. Both series also include a variety of isotopes of other nuclides. Since radium and thorium, and also some of the other elements, exhibit quite different behaviors in the organism, each transmutation of an isotope of one to that of another results in at least a potential translocation. Similarly, transmutation results in differences in radiation characteristics, including types of emanation, half-lives, and energies.

Translocation may be either into or away from bone. The decay of $Ra^{226}$ to $Rn^{222}$ permits diffusion of most of the radon (radium emanation) from the bone mineral and hence its escape from the body in expired air. The decay of $Th^{228}$ fixed in soft tissue—this element is associated with the organic matrix and mineral surfaces—to the bone-seeking $Ra^{224}$ is equivalent to a continuous infusion of radium into the blood stream from which it may be

deposited in bone mineral in the same manner as calcium. In the decay of $Th^{228}$, not all of the daughter products, however, are liberated into the blood stream; some remain behind in the soft tissues.

## 3.2. The Uranium-Radium Series

Table 3.2 lists the elements in the uranium-radium series, including the pathways of decay, the half-lives of the elements formed during decay, and the energies associated with the emission of alpha and beta particles. This series begins with $U^{238}$, which contributes 99.28% of the natural element and has a half-life of 4,500,000,000 years. Biological interest in this series begins with the most stable isotope of radium, $Ra^{226}$, which has a half-life of 1622 years.

$Ra^{226}$ decays by alpha-particle emission to the isotope of the noble gas radon, $Rn^{222}$, which has a half-life of 3.8 days. The radioactivity of radium derives primarily from the alpha particles emitted from $Ra^{226}$, from $Rn^{222}$, and from the daughter elements $Po^{218}$ and $Po^{214}$. Radium belongs to the alkaline earth series and apparently behaves much like calcium in the bone mineral. Radium is retained by the bone mineral for the lifetime of the individual except as it may be liberated by decay to radon, by solubilization and resorption of the mineral, or by long-term exchange. Long-term exchange, with excretion of the radium thus liberated, is the most important pathway for elimination of the element from the organism. Approximately 70% of the radon formed by decay of $Ra^{226}$ escapes from bone and is lost through the expired air. The other 30% remains in bone where each atom gives rise to three alpha particles, including those from its short-lived daughter elements. Consequently about one-half of the alpha particles from the radium and its daughters is emitted within the bone, and the other half is emitted by the radon (and its daughters) that escapes from bone. In the chain of events by which $Ra^{226}$ finally reaches the stable state of $Pb^{206}$, there is no translocation except for the loss of $Rn^{222}$ that significantly affects its biological effects; there is of course radiotoxicity at each stage of $\beta^-$ and $\alpha$ decay.

## 3.3. The Thorium Series

Thorotrast, a suspension of $Th^{232}$ dioxide, was introduced into clinical roentgenology in 1928[15] and was widely used during the period 1930–1945, mainly for visualization of the liver, spleen, and cerebral arteries. Its use depended upon its being radiopaque, insoluble, and of low toxicity. When administered intravenously, the colloidal particles were phagocytosed by macrophages, principally in the liver and spleen, where they remained fixed for long periods of time. When the radiotoxic hazard of thorotrast was rec-

TABLE 3.2

URANIUM-RADIUM SERIES

| Element Number | Isotope | Half-life | Energy (Mev) | Alternate Symbol |
|---|---|---|---|---|
| 92 | $U^{238}$ | $4.5 \times 10^{9}$ y | 4.18 ($\alpha$) | $U_I$ |
| | ↓ $\alpha, \gamma$ | | | |
| 90 | $Th^{234}$ | 24.1 d | 0.19 ($\beta^-$) | $UX_1$ |
| | ↓ $\beta^-, \gamma$ | | | |
| 91 | $Pa^{234m}$ | 1.14 m | | |
| | $Pa^{234}$ | 6.7 h | 0.45 ($\beta^-$) | $UX_2$ |
| | ↓ $\beta^-, \gamma$ | | | |
| 92 | $U^{234}$ | $2.5 \times 10^{5}$ y | 4.76 ($\alpha$) | $U_{II}$ |
| | ↓ $\alpha, \gamma$ | | | |
| 90 | $Th^{230}$ | $8.0 \times 10^{4}$ y | 4.68 ($\alpha$) | $I_0$ |
| | ↓ $\alpha, \gamma$ | | | |
| 88 | $Ra^{226}$ | 1622 y | 4.78 ($\alpha$) | — |
| | ↓ $\alpha, \gamma$ | | | |
| 86 | $Rn^{222}$ | 3.8 d | 5.49 ($\alpha$) | — |
| | ↓ $\alpha, \gamma$ | | | |
| 84 | $Po^{218}$ | 3.05 m | 6.0 ($\alpha$) | RaA |
| | ↓ $\alpha$ | | | |
| 82 | $Pb^{214}$ | 26.8 m | 0.67 ($\beta^-$) | RaB |
| | ↓ $\beta^-, \gamma$ | | | |
| 83 | $Bi^{214}$ | 19.7 m | 0.4 ($\beta^-$) | RaC |
| | ↓ $\beta^-, \gamma$ | | | |
| 84 | $Po^{214}$ | $3.04 \times 10^{-7}$ s | 7.68 ($\alpha$) | RaC' |
| | ↓ $\alpha$ | | | |
| 82 | $Pb^{210}$ | 19.4 y | 0.02 ($\beta^-$) | RaD |
| | ↓ $\beta^-, \gamma$ | | | |
| 83 | $Bi^{210}$ | 5.0 d | 1.15 ($\beta^-$) | RaE |
| | ↓ $\beta^-, \gamma$ | | | |
| 84 | $Pb^{210}$ | 138.4 d | 5.3 ($\alpha$) | RaF |
| | ↓ $\alpha, \gamma$ | | | |
| 82 | $Pb^{206}$ | Stable | — | RaG |

TABLE 3.3
THORIUM SERIES

| Element Number | Isotope | | Half-life | Energy (Mev) | Alternate Symbol |
|---|---|---|---|---|---|
| 90 | $Th^{232}$ | | $1.4 \times 10^{10}$ y | 4.00 ($\alpha$) | |
| | ↓ | $\alpha$, $\gamma$ | | | |
| 88 | $Ra^{228}$ | | 6.7 y | 0.012 ($\beta^-$) | $MsTh_1$ |
| | ↓ | $\beta^-$, $\gamma$ | | | |
| 89 | $Ac^{228}$ | | 6.1 h | 2.18 ($\beta^-$) | $MsTh_2$ |
| | ↓ | $\beta^-$, $\gamma$ | | | |
| 90 | $Th^{228}$ | | 1.9 y | 5.42 ($\alpha$) | RdTh |
| | ↓ | $\alpha$, $\gamma$ | | | |
| 88 | $Ra^{224}$ | | 3.6 d | 5.68 ($\alpha$) | ThX |
| | ↓ | $\alpha$, $\gamma$ | | | |
| 86 | $Rn^{220}$ | | 51.5 s | 6.28 ($\alpha$) | Tn |
| | ↓ | $\alpha$, $\gamma$ | | | |
| 84 | $Po^{216}$ | | 0.16 s | 6.77 ($\alpha$) | ThA |
| | ↓ | $\alpha$ | | | |
| 82 | $Pb^{212}$ | | 10.6 h | 0.58 ($\beta^-$) | ThB |
| | ↓ | $\beta^-$, $\gamma$ | | | |
| 83 | $Bi^{212}$ | | 60.5 m | 6.08 ($\alpha$) | ThC |
| | ↓ | $\alpha$, $\beta^-$, $\gamma$ | | | |
| 81 | $Tl^{208}$ | | 3.1 m | 1.79 ($\beta^-$) | ThC'' |
| | ↓ | $\beta^-$, $\gamma$ | | | |
| 82 | $Pb^{208}$ | | Stable | — | — |

ognized, its use fell into disfavor, but in the meantime a large number of individuals had been given significant body burdens of radioactivity. It is the purpose of this section to review the chain of events set in motion by the intravenous use of thorotrast. Thorium salts were used also for local injection into muscles or connective tissue where they were retained at the site of injection and only slowly released to the circulation. For practical

purposes it may be said that, once within the body, thorium is retained indefinitely.

The decay of thorium, beginning with $Th^{232}$ which has a half-life of 1.39 $\times 10^{10}$ years and ending with stable $Pb^{208}$, is shown in Table 3.3. Biological interest in this series begins with $Th^{232}$ itself, which may induce tumors of the soft tissues in which it is deposited. To the biological effects of $Th^{232}$ per se must be added those that may derive from the daughters—$Ra^{228}$ (mesothorium$_1$), which results from alpha decay of $Th^{232}$ and has a half-life of 5.85 years, and $Th^{228}$ (radiothorium), a product of $\beta^-$ decay of $Ra^{228}$ through the short-lived (6.1 hours) intermediate $Ac^{228}$ (mesothorium$_2$). Thorium-228 has a half-life of 1.9 years; it decays to $Ra^{224}$ (thorium X), which also has a short half-life (3.6 days). Radium-224, by alpha decay, yields an isotope of the noble gas radon ($Rn^{220}$, also called thoron) which has a half-life of 51.1 seconds. The energy released by $Ra^{228}$ is very small (see Table 3.3); most of the energy or radiation dose is carried by the short-lived daughters of $Ra^{224}$.

When thorotrast is injected intravenously, it is largely held in the macrophages of the liver and spleen. Approximately two-thirds of the $Ra^{228}$ derived from $Th^{232}$ is retained with the phagocytosed thorotrast, and about one-third enters the blood stream and is available for translocation, mainly to the skeleton where it behaves like $Ra^{226}$ and deposits in the bone mineral like calcium.[16,17] Van Dilla *et al.*[18] administered $Th^{228}$, in equilibrium with its daughters, intravenously to beagles. Thorium-228 behaves much like plutonium and is tightly bound to the organic matrix and/or mineral surfaces of bone. Its daughter $Ra^{224}$ enters the circulation and is then deposited as though it were radium injected intravenously; this was demonstrated by means of autoradiographic studies of bone removed at autopsy. The net result is a translocation of $Ra^{224}$ to bone mineral *in vivo*. Radium-228 has also been found in the bones of thorotrast patients and is believed to have been translocated from the $Th^{232}$ in the soft tissues. Radium-224 decays to $Rn^{220}$, which because of its short half-life has little opportunity to escape from the bone mineral and from the organism. Further decay in the thorium series is by way of short-lived elements, terminating with the stable $Pb^{208}$. There is little biological interest beyond $Rn^{220}$.

On the basis of available information,[16] a person who has carried a body burden of 4 g of $Th^{232}$ as thorotrast for several years should have the following range of radioactivity (in curies) in the skeleton:

*Minimum:*

$$1.8 \times 10^{-9}\,Ra^{228} + 1.8 \times 10^{-9}\,Th^{228} + 1.9 \times 10^{-8}\,Ra^{224} + 1.7 \times 10^{-8}\,Rn^{220}$$

*Maximum:*

$$1.8 \times 10^{-9}\,Ra^{228} + 5.5 \times 10^{-8}\,Th^{228} + 3.7 \times 10^{-8}\,Ra^{224} + 3.3 \times 10^{-8}\,Rn^{220}$$

Again, as in the case of cesium-137, decay of thorium-232 results in the formation of radioisotopes that leave the soft tissues for translocation in bone. An investigation of the late effects of thorotrast has been carried out by Looney,[15] who determined the status of 35 patients who had received thorotrast up to 23 years before the study was made. In a total of more than 5000 patients who had been given thorotrast, 10 or possibly 11 tumors were found, almost all in the liver; no tumors of bone were reported. Looney suggests that hepatic tumors will emerge as the predominant tumor in thorotrast patients just as bone tumors had become predominant in luminous dial workers and patients who received radium medically.

Faber[19] has also reported on a group of 842 patients who had been injected with thorotrast and whose case histories had been followed in Denmark for more than 14 years. This author states that the literature does not so far contain a single case of osseous or periosteal tumor in a patient injected with thorotrast and that no such case has been observed in the Danish patients mentioned above. Faber comments that the gradual way in which $Ra^{228}$ is produced would, according to his expectations, preclude the production of bone cancer.

Since it has been shown[18] that $Ra^{228}$, when administered in adequate doses, is capable of producing osteogenic sarcoma, it must be concluded that the failure of the radium daughters of $Th^{232}$ to induce malignancy in bone must be due to one of three possibilities: (1) that not enough time has elapsed for the appearance of malignant tumors of skeletal origin; (2) that the skeletal dosage of the radioisotopes of radium in the thorium series, following retention of thorotrast in the body, is not sufficient to result in osteogenic sarcoma; or (3) that, while bone sarcoma may have occurred, its incidence may have been so small that it has not been recognized in the patients studied.

## References

1. P. S. Chen, Jr., A. R. Terepka, and H. C. Hodge, The Pharmacology and Toxicology of the Bone Seekers, *Ann. Rev. Pharmacol.*, **1**: 369-396 (1961).
2. J. Duckworth and R. Hill, The Storage of Elements in the Skeleton, *Nutr. Abstr. Rev.*, **23**: 1-17 (1953).
3. R. C. Thompson, Vertebrate Radiobiology: Metabolism of Internal Emitters, *Ann. Rev. Nucl. Sci.*, **10**: 531-560 (1960).
4. P. W. Durbin, Distribution of the Transuranic Elements in Mammals, *Health Phys.*, **8**: 665-671 (1962).
5. J. Vaughan, Bone Disease Induced by Radiation, *Intern. Rev. Exptl. Pathol.*, **1**: 243-369 (1962).
6. C. P. Leblond, L. F. Bélanger, and R. C. Greulich, Formation of Bones and Teeth as Visualized by Radioautography, *Ann. N. Y. Acad. Sci.*, **60**: 631-659 (1955).
7. C. W. Asling and L. S. Hurley, The Influence of Trace Elements on the Skeleton, *Clin. Orthopaed.*, **27**: 213-264 (1963).

8. I. H. Tipton, W. D. Foland, F. C. Bobb, and W. C. McCorkle, *Spectrographic Determination of Trace Elements in Human Tissue,* USAEC Report CF-53-8-4, Oak Ridge National Laboratory, March 11, 1953.
9. F. C. McLean and M. R. Urist, *Bone: An Introduction to the Physiology of Skeletal Tissue,* 2nd ed., p. 153, The University of Chicago Press, Chicago, 1961.
10. D. Strominger, J. M. Hollander, and G. T. Seaborg, Table of Isotopes, *Rev. Mod. Phys.,* **30**: 585-904 (1958).
11. G. S. Stent and C. R. Fuerst, Genetic and Physiological Effects of the Decay of Incorporated Radioactive Phosphorus in Bacterial Viruses and Bacteria, *Advan. Biol. Med. Phys.,* **7**: 1-75 (1960).
12. P. W. Durbin, M. H. Williams, M. Gee, R. H. Newman, and J. G. Hamilton, Metabolism of the Lanthanons in the Rat, *Proc. Soc. Exptl. Biol. Med.,* **91**: 78-85 (1956).
13. D. L. Buchanan and A. Nakao, Studies on the Nature of Bone Carbonate, *Arch. Biochem. Biophys.,* **77**: 168-180 (1958).
14. J. R. Totter, M. R. Zelle, and H. Hollister, Hazard to Man of Carbon-14, *Science,* **128**: 1490-1495 (1958).
15. W. B. Looney, An Investigation of the Late Clinical Findings Following Thorotrast (Thorium Dioxide) Administration, *Am. J. Roentgenol. Radium Therapy Nucl. Med.,* **83**: 163-185 (1960).
16. L. D. Marinelli, *Retention and Elimination of Radium Isotopes Produced by the Decay of Thorium Parents Within the Body. E. Comparison with Experiment,* USAEC Report ANL-5689, pp. 29-35, Argonne National Laboratory, November 1957.
17. L. D. Marinelli and H. F. Lucas, Jr., Translocation of Thorium Daughters to Bone, in *Some Aspects of Internal Irradiation,* pp. 499-516, Pergamon Press, London, 1962.
18. M. A. Van Dilla, B. J. Stover, and J. S. Arnold, On the Retention and Translocation of $Ra^{224}$ (Th X) in Dogs, *Am. J. Roentgenol. Radium Therapy Nucl. Med.,* **77**: 503-510 (1957).
19. M. Faber, Thorotrast in Man—the Carrier State and the Sequelae, in *Some Aspects of Internal Irradiation,* pp. 473-498, Pergamon Press, London, 1962.

# Chapter 4

# *Detection and Measurement of Radioactivity*

## 1. Ionization and Excitation

Radiation, whether ionizing radiation or visible light, acts on the electronic structure of atoms. If the radiation, either particulate or electromagnetic, is of sufficient energy, it may cause the entire removal of an orbital electron from the atom. This is called *ionization*. The subsequent history of the detached electron and of the parent atom may assume one of several forms and will be discussed below.

If the energy of the radiation absorbed by an atom is not sufficient to result in ejection of an electron, it may result in a perturbation of the otherwise stable interrelations between the nucleus and the orbital electrons. This phenomenon is called *excitation*. If the electron continues to absorb energy from repeated exposure to radiation, it may be excited sufficiently to break the bonds of stability holding it in balance and to jump free of the atom. This separation of an electron from the atom marks the boundary between excitation and ionization, both of which are essentially of the same nature. The two conditions differ only in that the accumulated energy necessary to result in ionization is greater than that resulting in excitation alone.

Excitation of an atom, under the influence of radiation, commonly manifests itself in the emission of electromagnetic radiation during which the excited atom seeks a level of stability. An alternate method of loss of excess energy is by direct transfer of excitation energy to an adjacent atom.

Either excitation or ionization may be induced in an atom by particulate radiation, commonly by alpha or beta particles or by electromagnetic radiation; either particulate or electromagnetic radiation capable of inducing ionization is designated as *ionizing radiation*. For electromagnetic radiation, the limits of ionizing radiation are within the spectrum of ultraviolet light. Energetic ultraviolet light, with energies of 7 to 12 ev, corresponds to wavelengths between 1700 and 100 A and is capable of producing ionization; these wavelengths are immediately longer than those of X-rays (which also produce ionization) and shorter than ultraviolet radiation near the visible spectrum.

## 2. Nuclear Radiation Detectors

The quantitative determination of radioactivity depends on the *ionization* or *excitation* of matter by the radiations emitted by radioactive elements or by other sources of electromagnetic radiation. It is apparent that the usefulness of any radioisotope as a tracer lies in the detection, recording, and measurement to provide a system for such quantitative determination. Although significant advances in the development of instruments have been made, not all are applicable to the study of bone and teeth. Selection of the method to be used depends upon technical requirements as well as upon financial considerations. The scope of this volume permits only a brief description of the basic principles of the systems limited to tracer techniques and excludes those used in radiation therapy and dosimetry. Adequate treatment of instrumentation, including characteristics of counting, can be obtained from references on methodology.[1-8]

For practical purposes, the nuclear-radiation detectors may be divided into two general categories[9]: (1) *pulse systems,* in which the output of the detector represents individual interactions of a nuclear particle or a gamma ray with the detector; and (2) *non-pulse detection systems,* in which no attempt is made to resolve the individual interactions of radiation, so that the output represents the average or total cumulative effects.

## 3. Pulse Systems

Pulse systems may be further divided into: (1) those using the principle of ionization and ion collection, such as ionization chambers, proportional and Geiger-Müller tube counters; and (2) those using the excitation principle, such as the scintillation detectors.

### 3.1. Use of Ionization Principle

All ion-collection detectors have the basic features of an electrode system consisting of a circuit through which the current passes when ions, collected in a gas medium, are attracted to the charged electrodes.[1] The cathode and anode are connected through a resistance which permits a potential difference to be applied between them. The resistance may be inserted into the circuit or it may be the parts of the circuit.

When a charged particle passes through the gas, electrostatic interaction occurs between the particle and the orbital electrons of the gas molecule. Ion pair production results from the splitting of the molecules of gas along the path of the particle. The number of ion pairs formed per unit path length is the *specific ionization* and depends on the charge and speed of the particle. The electrical charge collected after passage of the particles varies owing to

the different numbers of ions formed by them as they traverse the gas in the collecting system. For example, alpha particles, being doubly charged and relatively slow moving, will produce a specific ionization greater than that of beta particles. Gamma rays and X-rays do not ionize gases directly since they are not themselves charged. However, ionization occurs by the secondary beta particles produced by their absorption, either in the gas itself or in the walls of the counting vessel.

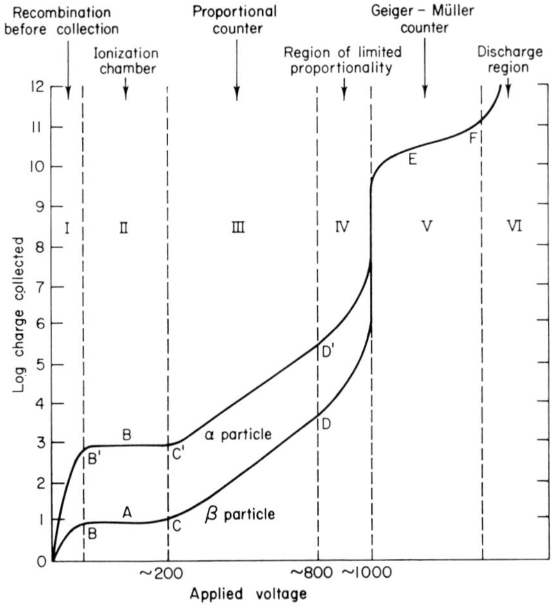


FIG. 4.1. Curves of pulse height vs. applied voltage to illustrate ionization, proportional, and Geiger-Müller regions of operation. (Adapted from Overman and Clark, *Radioisotope Techniques*.[1])

The differences in types of instruments depend primarily on the magnitude of the voltage applied to their electrodes and are best described in terms of the region of operation, as illustrated in Fig. 4.1.

### *3.1.1. Ionization Chamber Region*

In region II, ion chambers can be operated as pulse systems by counting individual events.[1] Ionization chambers can also be operated as non-pulse detection systems, i.e., as current devices. These devices will be described in a later section.

At low voltage a very small fraction of ions formed in the detector can be

collected but with increase in potential the field will increase proportionately so that all the ions produced inside the detector are collected. With increasing voltage, more ions are collected before they are able to recombine; this proportion increases until almost all ion pairs that are produced can be collected. A further increase in voltage will have no appreciable effect on the collection efficiency nor on the current in the circuit; this voltage range is the *saturation voltage*, or *saturation plateau*. It is in this region that an ionization chamber is normally operated.

### *3.1.2. Proportional Region*

When counter tubes are operated in this region,[1] it merely indicates that the size of the electric pulse (output pulse) is proportional to the number of electrons released by the particles. All primary ion pairs produced are being collected at the electrodes. With higher potentials, some of the primary electrons are accelerated in the field of the electrode to such a point that they interact with other gas molecules, thereby producing secondary ionization. *Gas amplification* occurs when more ions reach the collecting electrode than were in the original collisions and the charge collected increases over the saturation value. As indicated by CD and C′D′ in Fig. 4.1, the increase in the charge collected at the electrode is proportional to the voltage. The number of ions collected will remain proportionately different for alpha and beta particles, owing to the different number of primary ions formed by the radiation. The discharge is localized to a small portion of the tube in this region.

### *3.1.3. Geiger Region*

In region V the charge collected is independent of the alpha and beta radiation and the number of primary ions initially formed—all ionization, regardless of its nature, will produce similar end results.[1] The detector used in this high-voltage region is the Geiger-Müller tube counter. The tube itself is a gas-filled (helium or argon) cylindrical cathode and an axial wire anode. A high electric voltage, approximately 1000 volts, is applied between the electrodes. The field intensity about the axial electrode is so high that any ion formed, whether of primary or secondary origin, can be accelerated enough to cause additional ionization of the gas molecules. A chain reaction is created, and an avalanche of electrons and positive ions will form unless the tube is quenched. Most tubes used in biology are self-quenching; i.e., they are filled with a mixture of a noble gas and an organic vapor. The discharge is stopped after a fraction of a second, and the tube is then prepared to respond to another impulse; this interval is the dead time and, for a typical Geiger-Müller tube, is of the order of 200 to 300 μsec.

### 3.2. Use of Excitation Principle

When electromagnetic (gamma rays or X-rays) radiation interacts with matter,[1] ionization is also produced, but it occurs by any of the three possibilities illustrated in Fig. 4.2. These are: the *photoelectric effect*, the *Compton interaction* (Compton scattering), and *pair production*. The method by which ionization occurs by any of the possibilities depends on the energy and velocity of the charged electron and the atomic weight of the matter (absorber).

In the photoelectric effect where there is low energy and the absorber is of high atomic weight, all of the energy of a gamma-ray photon is transferred to the ejected electron. In the Compton interaction, only part of the energy of a medium-energy gamma-ray photon is transferred to the electron. This interaction will occur in absorbers of any atomic weight. In pair production, the high-gamma-ray energy goes into the formation of a positron and a negatron and imparts kinetic energy to them; this occurs in absorbers of high atomic weight. These electrons then lose energy by ionization and excitation until they are stopped. The positron then combines with an electron and both particles are totally converted to annihilation quanta or gamma rays of electromagnetic energy.

There are many chemical substances which absorb excitation energy and re-emit the energy in the form of light photons. Such substances are called *phosphors* or *scintillators*. When radiation interacts with phosphors, kinetic energy of particles or of electromagnetic radiation is converted into flashes of visible or ultraviolet light. The total amount of light emitted is proportional to the energy absorbed by the phosphor. The light emitted is close to the limit of the visible spectrum, i.e., circa 4000 A (see Table 4.1). This conversion of energy to flashes of light is *scintillation*.[1,11,12]

#### 3.2.1. *The Scintillation Detector*

The principle of the scintillation detector is that of translating radioactivity to quanta of light and then recording the pulses by a photomultiplier tube. In general the basic system consists of a detector (crystal and photomultiplier), a high voltage supply, a preamplifier, and a scaler.

The properties of scintillation detectors depend on the type of phosphor used, i.e., solid or liquid. Solid phosphors are either inorganic crystals or organic crystals; liquid phosphors are either organic liquids or plastics. The most common inorganic crystal in use for gamma-ray detection is thallium-activated sodium iodide, usually referred to as NaI(Tl) crystals. This substance has the highest light output per unit of dissipated energy of all known phosphors. This property together with the relatively large photoelectric cross section of the crystal gives it good energy resolution of X-rays. The organic

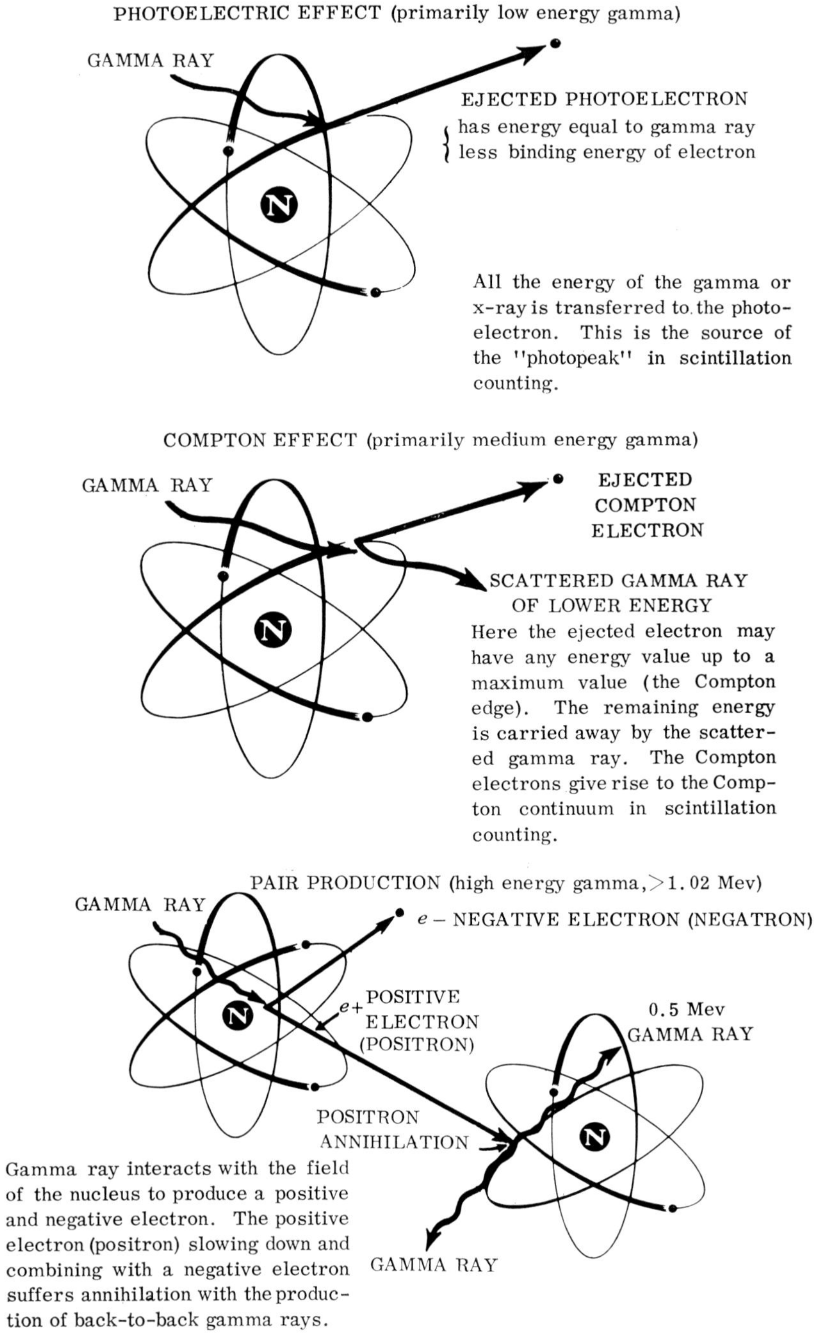


FIG. 4.2. Ionization by excitation. (From *Scintillation Spectrometry*.[14])

crystals have a smaller light yield and a faster pulse than the inorganic crystals. They are of particular value for beta-ray detection, especially as a thin crystal which has low gamma sensitivity. An example is anthracene, with approximately one-half the light yield of NaI and one-tenth the pulse length. Organic liquids are used for assay of soft beta-emitting isotopes dissolved in the liquid. An example is diphenyloxazole dissolved in toluene. It has a light yield of approximately one-half and a pulse length one-third that of anthracene. An example of a plastic scintillator is terphenyl in polystyrene. The light yield is approximately one-third that of anthracene and the pulse length is one-fifth. Plastic scintillators (as compared with NaI crystals) are relatively inexpensive and may be produced in various shapes and in large sizes; these are their chief advantages. The characteristics of various scintillators are shown in Table 4.1.

When ionizing radiation strikes a phosphor, electrons are released from their atoms. Other less tightly bound electrons shift into the vacated orbits and cause an emission of light whose energy is equal to the differences in binding energy of the initial and final stages of the electrons. The light may escape from the phosphor or may be absorbed by another electron in the phosphor. The absorbing electron may then escape from its molecular orbit or rise to an excited state. Another electron may, in turn, drop to this vacated orbit, and another photon of light of lower energy than the first photon may be emitted. For each ion pair formed by the initial ionizing particle, approximately one photon is emitted in the phosphor.

A light-reflection surface usually surrounds the phosphor so that the light escapes primarily from one surface only. A photomultiplier tube is placed at this surface. The inside surface of the end of the tube is coated with a very thin layer of a material with a high photoelectric absorption coefficient. This surface is called the *photocathode*. Electrons are released from this surface by the light from the phosphor. The kinetic energy of the electrons so released is only of the order of 3 to 4 ev, and therefore they cannot penetrate very far. As a result, only a very thin coating can be used for this conversion from photons to photoelectrons. The photoelectrons escaping from the photocathode surface are accelerated to the first dynode of a 10-stage cascade electron multiplier by a potential difference of about 150 volts. With this added kinetic energy, the electrons share their energy with some of the surface electrons that they strike. For each electron (or other ionizing particle) that loses energy in the phosphor, a discrete electrical pulse will appear at the anode of the photomultiplier tube. The pulse size will be proportional to the energy per particle lost in the phosphor and to the gain per dynode in the photomultiplier tube, since the gain is a function of the voltage across the tube.

The size of the electrical pulse produced by each gamma ray entering the

TABLE 4.1
PHYSICAL CONSTANTS OF SELECTED SCINTILLATION PHOSPHORS[a]

| Type | Density | Refractive index | Pulse height (% anthracene) | Decay period (mean life), seconds | Wavelength of maximum emission, A | Principal applications |
|---|---|---|---|---|---|---|
| Plastic | 1.04 | 1.581 | 40–70 | $4.0 \times 10^{-9}$ | 4300 | Gamma, alpha, beta, fast neutron |
| Liquid | 0.88 | 1.504 | 78 | $2.8 \times 10^{-9}$ | 3920 | Alpha, beta (internal counting, low temperature) |
| Gel | 0.88 | 1.496 | 50 | $2.6 \times 10^{-9}$ | 4300 | Alpha, beta (internal counting) |
| Pb (5%) | 0.93 | 1.505 | 25 | $1.1 \times 10^{-9}$ | 4300 | Gamma |
| Anthracene | 1.25 | — | 100 | $3.2 \times 10^{-8}$ | 4470 | Gamma, alpha, beta, fast neutron |
| Stilbene | 1.16 | 1.626 | 60 | $4.0 \times 10^{-9}$ | 4100 | Gamma, alpha, beta, fast neutron |
| NaI(Tl) | 3.67 | 1.775 | 230 | $0.25 \times 10^{-6}$ | 4130 | Gamma |
| ZnS(Ag) | 4.09 | 2.356 | 300 | $2 \times 10^{-7}$ | 4500 | Alpha, beta |

[a] From Overman and Clark, *Radioisotope Techniques*.[1]

TABLE 4.2

METHODS FOR SAMPLE COUNTING AND ASSAY[a]

| Emission | | Example | Radiation detectors |
|---|---|---|---|
| Very soft | β-emitters | $H^3$ | As a part of a solute in a liquid scintillator, or as part of a gas in a chamber or counter. |
| Soft | β-emitters | $C^{14}$, $S^{35}$ | As a gas either in a chamber or counter; thin solid samples with a sensitive beta-detector, e.g., a mica window G-M counter, very thin window flow counter, or a windowless flow counter. Liquid scintillator. |
| Medium | β-emitters | $Ca^{45}$ | Usually assayed as solids, otherwise same as $C^{14}$ and $S^{35}$ methods. |
| Hard | β-emitters | $P^{32}$, $Y^{90}$ | Beta-sensitive counters. Liquid samples in a pan or surrounding a thin-walled dip counter or solution counter. Liquid scintillator. |
| Medium | β-emitters with much γ-emission | $Mg^{28}$ | Beta-sensitive equipment can be used. For small volume samples a well-type scintillation counter is used. For large volume samples an annular, multiple-element G-M counter is available with a gamma-sensitivity comparable to that of an equivalent plastic or liquid phosphor. |
| Hard | β-emitters with little γ-emission | $Ba^{140}$ | Choice between a well-type scintillation counter or measured as beta-emitter. |
| Hard | β-emitters with much γ-emission | $Na^{24}$ | Well-type scintillation counter, or measured as beta-emitters. |
| Positron emitters | | $F^{18}$ | Most easily assayed by measuring annihilation radiation in a well-type scintillation counter. |
| EC isotopes | | $Be^7 \rightarrow Li^7$ | Special G-M or proportional counters with Be (low $Z$) windows and an absorbent (high $Z$) gas filling. Proportional counters with pulse-height analyzer for standardization and identification. |

[a] Derived from Robinson, *Medical Physics*.[11]

TABLE 4.3
DETECTORS AND ASSOCIATED EQUIPMENT[a]

| Equipment | Description |
|---|---|
| Ionization chambers | Chambers with separate and sensitive electrometers; with incorporated Lauritsen electroscopes; portable chamber and galvanometer arrangements. |
| Proportional Geiger-Müller counter tube | Both may be constructed as sealed tubes or as flow counters. Cylindrical tubes: thick-walled for gamma rays; thin-walled for gamma and beta rays. End window counters with mica windows. Windowless flow counters for soft beta emitters. |
| Voltage supply | High voltage supplied by a regulated electronic circuit. Batteries and transistors for portable equipment. Current drain on ionization chambers and counters usually negligible; that for photomultiplier tube depends on resistor stack supplying the dynodes. |
| Pulse amplifiers | For counting and for pulse-height analysis. Must have: fast rise time and short resolving time; provisions for non-overloading; gain stability. Linearity of response needed for pulse-height work. |
| Preamplifiers and cathode followers | All detectors are high-impedance devices and work best into a high-impedance input. Also where signal must be passed through a long or high capacitance (low impedance) cable. |
| Scaling circuits (scalers) | Pulses from a detector, either direct or via an amplifier may be fed into a scaler which in turn will drive a register at a reduced rate. Requirements are: short resolving time; proper input sensitivity; proper output pulse for driving the register; dependability. |
| Timers | Various degrees of automation; either pre-set count with time indicated or pre-set time with count indicated. |
| Registers | Commonly digital, rotary, tubular devices which can be driven by an electrical pulse from the scaling unit. Also used are dial-type registers with faster response. |
| Rate meters | Circuits indicate history of the counting rate, with a length of time of integration or time constant which can be set for different values. Rate meter-recorders for chromatogram scans and pulse-height analyzers. |
| Pulse height analyzers | In single channel type, signals received from a linear pulse amplifier. Analyzer produces a uniform output pulse for each pulse from the amplifier which lies in a certain voltage range setting of width equal the window and lower limit equal the base line. Output pulse drives a scaler of rate meter-recorder. Pulse-height spectrum obtained by keeping a constant window and scanning the base line at a constant rate. Multiple-channel analyzers keep nearly all pulses and sort them into channels; the more channels, the more accurate the presentation of the spectrum. |

TABLE 4.3 (*Continued*)

| Equipment | Description |
|---|---|
| Coincidence and anti-coincidence circuitry | Circuits have two inputs and an output. Coincidence circuit gives an output pulse for coincident input pulses arriving within its resolving time. The anti-coincidence circuit gives an output pulse for input pulses except when a coincidence occurs. Used in standardization, in low-level, low-background counting, in measuring gamma-gamma coincidences, and liquid scintillation counting. |
| Other equipment | Shielding of lead, iron or other material to reduce background; gas supplies for flow counters; sample changing equipment; print-out equipment; scanning and recording equipment for *in vivo* mapping of activity in the body; and vacuum equipment for handling gas samples. |

[a] Derived from Robinson, *Medical Physics*.[11]

phosphor is proportional to the energy dissipated by the gamma ray in the crystal. Consequently, as the voltage is increased, amplification of all the pulses is proportionately increased, and the largest pulses, representing the most energetic gamma rays, are detected first. As the voltage is further increased, amplificaton is increased with the result that smaller pulses, representing less energetic gamma rays, are detected and counted also. This continues until all pulses are detected and recorded. The result is a curve characterized by a series of steps each of which represents a particular photopeak produced either by the primary radiation, scattered radiation, or noise; this curve is called the *integral spectrum*. Each gamma-emitting isotope produces a characteristic curve which can be used for its identification. It is often more convenient to use a multichannel pulse-height analyzer where amplification may be kept constant.

### *3.2.2. Summary*

The choice of instruments applicable to tracer work in the study of hard tissues is largely governed by the specific radioisotope being used, the energy of the radiation, the dose administered, and the preparation of the tissue sample. Ionization chambers and proportional counters are satisfactory for the measurement of short-range alpha particles, although these instruments cannot be used as efficiently for particles having a range larger than the dimensions of the chambers. On the other hand, the phosphors used with scintillation detectors can absorb the secondary electrons produced by gamma or beta particles, thereby making measurements of the energies of these particles practicable. While each gas counter requires a tube somewhat different in construction and filled with a special gas, it is possible to operate the same tube as an ionization chamber, proportional counter, or Geiger counter by increasing the voltage. Scintillation-counting techniques are particularly useful in beta and gamma energy measurements.

Although selection of a detector involves a variety of technical requirements, the final decision usually rests on financial considerations. The various detectors possess different degrees of complexity. A Geiger counter tube drives a sensitive scaler directly; proportional counting requires the addition of a nonoverloading amplifier. A scintillation system operated only for sample counting utilizes the integral discriminator circuit in the amplifier or scaler while differential analysis requires both a nonoverloading amplifier and a pulse-height analyzer of one or more channels.

A complete discussion of all types of detectors and their uses will not be given here. Rather, examples of methods and applications are summarized in Table 4.2 and detectors with associated equipment are listed in Table 4.3.

## 4. Nonpulse Systems

Nonpulse systems can be divided into: (1) current-producing systems in which the output is proportional to the total ionization current produced within the detector and (2) the photographic system.

### 4.1. Current-Producing System

Measurements are made with ionization chambers with an integrating circuit. The chambers themselves are made in a variety of forms and are used for measuring gaseous or solid samples. They are connected to either a vacuum tube or a dynamic-capacitor electrometer circuit. The electrometer gives readings by either the rate-of-drift or the constant-deflection method. This technique is particularly useful in the measurement of $C^{14}$ in small bone samples where the tissue can be dissolved and counted in the form of $CO_2$ or a solid carbonate.

### 4.2. Photographic System

When a piece of film is exposed to radiation, the emulsion reacts in the same way as when struck by visible light. The amount of radiation can be estimated from darkening of the film. Film can be considered a nuclear-detecting device; it is used either in the form of a film badge to determine the accumulation of radiation or as a data-registering device—a localizing detector—as with autoradiograms. The components of any film or plate are (1) the sensitive silver halide material, (2) a dispersal medium for the sensitive material, usually gelatin, and (3) the emulsion base, which is cellulose acetate for most films and glass for slides.

#### *4.2.1. Film Badge*

For personnel monitoring, the density of the blackening on the emulsion may be compared with that of a calibrated film.

#### *4.2.2. Autoradiography*

The oldest method used for recording and crudely measuring radioactivity was that of fogging of photographic emulsions, used for this purpose as early as 1896 by Becquerel. This procedure has now been greatly refined to the extent that it may be used under certain conditions for quantitative measurements in addition to the usual qualitative demonstration of radioactivity by autoradiograms.

In its simplest terms, an autoradiogram is the image produced on a photographic emulsion by radioactivity emanating from the sample. Many variations in procedure and in materials are available, and the choice depends

upon the conditions of the observations to be made and upon the kind of records it is desired to produce. In general, following administration of a radioactive element to an organism, a tissue sample is fixed, dried, embedded, and sectioned. The section is then pressed into close contact with a photographic film or is coated with an emulsion stripped from a film or with a fluid emulsion. On subsequent development of the emulsion, all regions corresponding to localization of the radioactive element are darkened; prints may be enlarged to any desired size. The exposure time for the film or emulsion must usually be empirically determined.

Autoradiography is a powerful tool in the study of bone. For maximum usefulness, it is applied to undecalcified bone. Introduction of a method for sectioning undecalcified bone shortly before World War II has permitted many of the observations made on the distribution and behavior of bone-seeking radioisotopes.[13] The techniques used in autoradiography and interpretation of results are described in Chapter 10.

## References

1. R. T. Overman and H. M. Clark, *Radioisotope Techniques,* McGraw-Hill Book Co., Inc., New York, 1960.
2. C. L. Comar, *Radioisotopes in Biology and Agriculture,* McGraw-Hill Book Co., Inc., New York, 1955.
3. G. D. Chase, *Principles of Radioisotope Methodology,* Burgess Publishing Co., Minneapolis, 1959.
4. G. E. Francis, W. Mulligan and A. Wormall, *Isotopic Tracers,* 2nd ed., The Athlone Press, University of London, London, 1959.
5. E. H. Quimby, S. Feitelberg, and S. Silver, *Radioactive Isotopes in Clinical Practice,* Lea & Febiger, Philadelphia, 1958.
6. M. D. Kamen, *Isotopic Tracers in Biology,* 3rd ed., Academic Press Inc., New York, 1957.
7. E. R. King and T. G. Mitchell, *A Manual for Nuclear Medicine,* Charles C Thomas, Springfield, Illinois, 1961.
8. C. F. Behrens, The ionizing radiations, in *Atomic Medicine,* 3rd ed., pp. 76-92, The Williams & Wilkins Co., Baltimore, 1959.
9. *Radioisotopes in Science and Industry,* USAEC Special Report, January 1960.
10. *Atomic Radiation,* RCA Service Co., Camden, New Jersey, 1957.
11. C. V. Robinson, Isotopes: Radioactive; Detection and Measurement, Practical Aspects, in *Medical Physics,* Vol. III, pp. 308-314, The Year Book Publishers Inc., Chicago, 1960.
12. R. D. Evans and N. C. Rasmussen, Isotopes: Radioactive Measurement, in *Medical Physics,* Vol. III, pp. 338-341, The Year Book Publishers, Inc., Chicago, 1960.
13. F. C. McLean and W. Bloom, Calcification and Ossification. Calcification in Normal Growing Bone, *Anat. Record,* **78**: 333-359 (1940).
14. *Scintillation Spectrometry, A Handbook,* Baird-Atomic, Inc., Cambridge, Massachusetts, 1960.

# Chapter 5

# *The Mineral of Bones and Teeth*

## 1. Crystal Structure and Chemical Composition of the Bone Mineral

Most of the present knowledge concerning the chemical dynamics of the bone mineral and its surface chemistry is dependent upon isotopic methods. For orientation in the subject of the bone mineral, it is desirable to review briefly the current concepts of its structure and composition.

The mineral fraction of bone consists mainly of $Ca^{2+}$, $PO_4^{3-}$, $OH^-$, $CO_3^{2-}$, and citrate with small amounts of other ions, especially $Na^+$, $Mg^{2+}$, and $F^-$. The crystal structure and chemical composition of the bone mineral have been under investigation for more than 100 years, and they remain a subject of intensive study.

Certain statements may be made about the bone mineral[1] to serve as a guide to its composition, structure, and behavior: (1) The crystals are submicroscopic in size and of hardly more than colloidal dimensions. (2) The bone salt is not a single, homogeneous chemical individual. (3) The basic atomic structure is that of an apatite mineral, prototypes of which are common in nature. (4) The chemical composition is less clearly defined, partly because of its variability and partly because of the difficulty of differentiating between (a) the constituents of the basic structure, (b) the ions having only a surface relationship to the crystals, and (c) those possibly admixed with the apatite crystals in a separate phase. (5) The structure and composition of the mineral correspond most closely to that of hydroxyapatite in all vertebrates studied; the mineral component, however, includes carbonate and citrate in substantial amounts. (6) Various substitutions and exchanges of ions occur in the crystal structure. (7) Dissolution and recrystallization may occur. (8) The minute crystals and the voids and discontinuities in their structure expose large surfaces to the fluids of the body. (9) The ions held by adsorption or by substitution at these surfaces, as well as those incorporated within surface unit cells with unshared sides, have a considerable influence upon the total composition of the mineral. (10) The surface ions are subject to rapid exchange with the fluids bathing the crystals. (11) The reactivity of the mineral varies with its age, with its immediate environment, and with its structure and chemical composition.

## 2. Crystals of Bone Mineral

The crystals of the mineral of bone are extremely minute—too small to be seen with the light microscope. While it was first shown by X-ray scatter patterns that the crystals are of the order of a few hundred angstroms in size, current knowledge of the size and shape of the individual crystals is dependent mainly upon electron microscopy. This method has its limitations, but the

FIG. 5.1. Electronmicrograph in a region of loosely packed fibrils in bone. (From original by Glimcher, *Calcification in Biological Systems*.[5]) (Magnification × 115,000.)

resolving power of the electron microscope has been increased, and the crystal of bone mineral is within the limits of its resolution. On the basis of X-ray diffraction[2] and electron microscopic data[3] the crystals may be visualized as rodlike (with diameters of about 50 A) and composed of chains of microcrystals arranged end to end.[4] The chains are of indeterminate length, oriented in the long axis of the collagen fibrils and possessing a special relationship to their cross-banding. The possibility that many of the crystals are prismlike, with the smallest dimension of the order of 20–30 A, is not excluded. The size and shape of these crystals have a marked influence upon their surface

chemistry, upon the exchangeability of the surface ions, and possibly upon the ease with which they go into solution (Fig. 5.1).

It is convenient to consider the structure of the bone mineral as consisting of a large number of similar, but not identical, complexes. These may be regarded as including interior unit cells wholly within the crystals; surface unit cells, sharing some of their sides with interior cells; ions bound by surface forces; and a surface hydration shell, containing ions in equilibrium both with the surrounding medium and with the surface (Fig. 5.2).

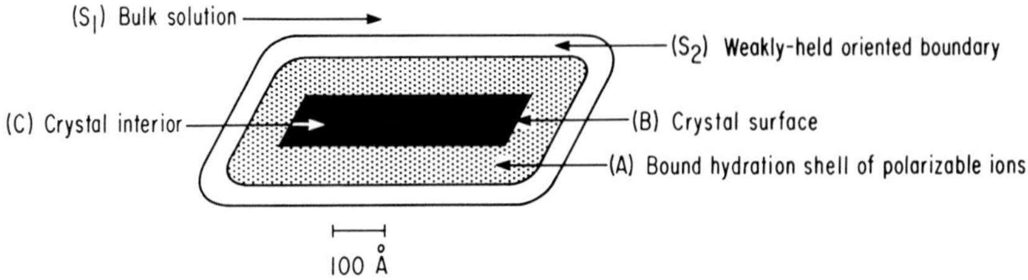


FIG. 5.2. A diagrammatic representation of a cross-section view of hydroxyapatite crystal in aqueous suspension. (From Neuman and Neuman, *The Chemical Dynamics of Bone Mineral.*[6])

## 2.1. Internal Structure of Crystals

Much of the present knowledge of the internal structure of the microcrystals of the bone mineral stems from interpretation of the X-ray diffraction pattern of the crystallites of bone as compared with patterns of relatively pure samples of synthetic or naturally occurring crystals of minerals of known composition. When a monochromatic X-ray beam is passed through a regular array of atoms (e.g., a crystal) a diffraction pattern is produced that is related to the atoms in the material and more especially to their spatial arrangement. By the use of a film near the irradiated sample, it is possible to measure the angles between the incoming X-ray beam and the outgoing diffracted beams. Data may thus be obtained concerning the orientation of the crystals.[7]

The mineral constituent of bone consists chiefly of crystals of a compound of calcium and phosphate with the structure of an apatite. The apatite series contains a number of compounds with their constituent ions arranged similarly in a three-dimensional symmetry pattern. The apatites usually have: (1) divalent cations ($Ca^{2+}$, $Pb^{2+}$, $Sr^{2+}$), (2) tetrahedral anionic radicals ($PO_4{}^{3-}$, $SiO_4{}^{3-}$), and (3) monovalent anions ($OH^-$, $F^-$), all in a hexagonal symmetry array characteristic of this isomorphous series. Certain ions can be substituted one for the other (e.g., $Pb^{2+}$ for $Ca^{2+}$, $F^-$ for $OH^-$) without disturbing the symmetry of the structure, although the distance between ions

may change owing to the differences in ionic size. X-ray diffraction techniques can easily detect changes of the order of 0.5% in interatomic differences resulting from isomorphous substitution.

X-ray diffraction, together with chemical analysis, has brought about general agreement that the basic structure of bone mineral is some form of hydroxyapatite,[1] of which the prototype is:

$$Ca_{10}(PO_4)_6(OH)_2 \qquad \text{(hydroxyapatite)}$$

It is convenient to examine first the internal structural characteristics of synthetic or other known apatites and then to compare the mineral of bones and teeth with these in an effort to arrive at a more precise understanding of the minerals as found in the animal body.

## 2.2. Synthetic Apatite Calcium Phosphates

The formulation of hydroxyapatite calls for a molar Ca/P ratio of 1.667. By mixing phosphoric acid and calcium hydroxide in various proportions, one may obtain calcium phosphates with Ca/P ratios as low as 1.5—all of these phosphates give the X-ray diffraction pattern of apatite. In this way, a continuous series of phosphates containing from 10 to 9 $Ca^{2+}$ to 6 $PO_4^{3-}$ may be prepared. The compound with the Ca/P ratio 1.5 has frequently been referred to as hydrated tricalcium phosphate[8] or as alpha tricalcium phosphate,[9] and has the following formulation:

$$3Ca_3(PO_4)_2 \cdot H_2(OH)_2 \qquad \text{(hydrated tricalcium phosphate; alpha tricalcium phosphate)}$$

Posner *et al.*[10] modified this view and considered that calcium ions are replaced by $H_2$ in the continuous series of apatite phosphates. Later, Posner and Perloff[11] suggested the formulation:

$$Ca_{10-x}H_{2x}(PO_4)_6(OH)_2 \qquad \text{(calcium-deficient hydroxyapatite)}$$

and indicated that Dallemagne's concept of alpha tricalcium phosphate is a special case of a calcium-deficient hydroxyapatite. Posner *et al.*[12] proposed that hydrogen bonding is responsible for maintaining neutrality in low-calcium hydroxyapatites, since hydrogen bonds are present in a series of such compounds in proportion to the missing calcium ions.

Hydrogen bonding in these compounds has been further demonstrated by Winand[13] in Dallemagne's laboratory, who proposed the following formula for the calcium-deficient apatite containing 9 $Ca^{2+}$ ions for 6 $PO_4^{3-}$ groups:

$$Ca_9(PO_4)_4(OH)\ (O_3PO{-}H{-\!-}OPO_3) \qquad \text{(hydrogen bonding in Ca-deficient hydroxyapatite)}$$

This hydrogen bonding, according to Winand, is partly or completely replaced by $Ca^{2+}$ for the compounds with higher Ca/P ratios. When the com-

pound is heated to 500–600°C, dehydration occurs, and the unit cells lose one molecule of $H_2O$ and give rise to a $P_2O_7$ group according to the formula:

$$Ca_9(PO_4)_5(OH)(P_2O_7)(OH)(PO_4)_5Ca_9 \qquad \text{(calcium pyrophosphate)}$$

Between 600°C and 700°C, a second dehydration occurs and pyrophosphate groups revert to orthophosphate; the beta form of tricalcium phosphate—$Ca_3(PO_4)_2$—is formed and is completely dehydrated.

When the apatite calcium phosphate contains $9\,Ca^{2+}/6\,PO_4^{3-}$, 16.6% of P is detectable as pyrophosphate. The proportion of pyrophosphate decreases linearly when the Ca/P ratio of the compound increases.[14] Hydroxyapatite ($10\,Ca/6\,PO_4$) does not contain any pyrophosphate.

Calcium-deficient apatites with still lower Ca/P ratios may be prepared; the lowest member of the series is octocalcium phosphate which contains 33.3% P as pyrophosphate after heating.[13] Calcium-deficient apatites with Ca/P ratios between 1.5 and 1.33 are solid solutions of octocalcium phosphate and higher members of the series.

### 2.3. The Unit Cell

The internal structure of the hydroxyapatite crystal is best understood in terms of the *unit cell.* This is a conceptual configuration and has no independent existence. It is the smallest expression of the ions found in the same ratios and in the same spatial relationships in which they are present throughout the entire crystal. Imaginary lines connecting the ions, drawn in an arbitrary but regular way, outline the unit cell; when extended through the crystal structure, these lines result in a three-dimensional lattice—the *crystal lattice.*

As shown in Fig. 5.3, the crystal structure may be viewed in cross-section as a series of rhombi, or equilateral parallelograms, with sides, or *a* axes, approximately 9.43 A in length. When extended in space, the third dimension, or *c* axis, oriented in the long dimension of the crystal of hydroxyapatite, measures approximately 6.88 A. Each of these three-dimensional geometric figures, in the form of a parallelepiped, is by convention a unit cell of the crystal lattice. Another method of visualizing the crystal structure in cross section is as a series of hexagons, representing a honeycomb arrangement of ions.

### 2.4. Chemical Nature of Bone Mineral

It has been noted above that the bone salt is not a single homogeneous compound, that the chemical composition is not uniform, that various substitutions and exchanges of ions occur in the structure, that dissolution and recrystallization may occur, and that there are voids and discontinuities in

the crystals. All of these factors make for an unsatisfactory situation with respect to chemical characterization of the mineral, even though its basic structure, corresponding closely to that of hydroxyapatite, is well established.

Table 5.1 from Armstrong and Singer[15] shows the results of elementary analysis of a single sample of dry, fat-free, bovine cortical bone and may be regarded as representing the proportions of the chief constituents of bone. The uncorrected molar Ca/P ratio is 1.656, a figure that corresponds closely to the theoretical ratio of 1.667 for hydroxyapatite. If corrections are made

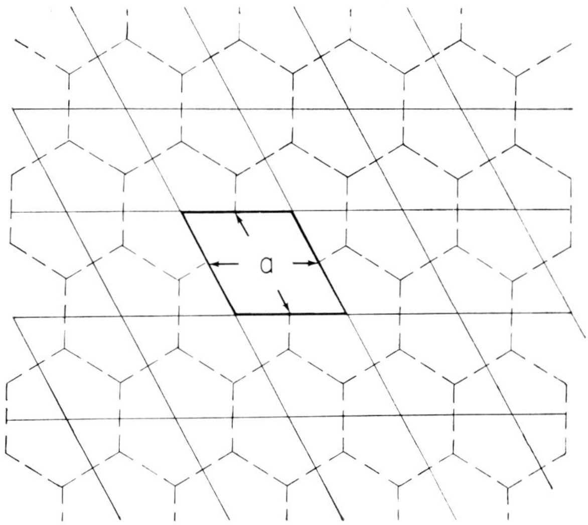


FIG. 5.3. Diagram of a cross section of a crystal of bone salt, to illustrate internal structure. The rhombi represent the unit cells of the crystal. The hexagons represent a honeycomblike arrangement of ions, oriented with the long axis of the crystal. At each intersection of the hexagons is a $Ca^{2+}$ ion. The tunnels which contain a column of $OH^-$ ions in the center are lined with $Ca^{2+}$ ions and packed with $PO_4^{3-}$ ions. (From McLean and Urist, *Bone: An Introduction to the Physiology of Skeletal Tissue*.[1])

for the $Ca^{2+}$ required to neutralize anionic constituents of the mineral (e.g., for carbonate, or for cations other than $Ca^{2+}$ available for neutralization of these constituents), the Ca/P ratio will vary accordingly.

Bone heated at 325°C contains between 3 and 15% phosphorus as pyrophosphate, an amount which has been found in synthetic calcium-deficient apatites.[14,16-18] Such large differences in the pyrophosphate content depend mainly on the age of the material under study. For example, in young rat bone, the yield of pyrophosphate decreases with the increasing age of the animal.[17] The same observations have been made on teeth. Rabbit incisors (dentin and enamel), which are continually growing, contain a larger

amount of pyrophosphate after heating than heated molars of the same animal.[19] According to Dallemagne,[20] old mineralized structures mostly contain hydroxyapatite; on the other hand, the mineral part of young bone and teeth is an apatite calcium phosphate with a lower Ca/P ratio and contains hydrogen bonds corresponding to the content of pyrophosphate in the heated material. This calcium-deficient apatite approximates the compound frequently referred to as hydrated tricalcium phosphate; a compound which matures with aging and is transformed into hydroxyapatite in old bone and teeth. Moreover,

TABLE 5.1
COMPOSITION OF DRY, FAT-FREE, BOVINE CORTICAL BONE[a]

| | Per cent | mEq/g |
|---|---|---|
| *Cations:* | | |
| Calcium | 26.70 ± 0.02 | 13.32 ± 0.01 |
| Magnesium | 0.436 ± 0.009 | 0.358 ± 0.008 |
| Sodium | 0.731 ± 0.015 | 0.318 ± 0.007 |
| Potassium | 0.055 ± 0.0009 | 0.014 ± 0.0002 |
| Total | | 14.01 |
| *Anions:* | | |
| Phosphorus | 12.47 ± 0.013 | — |
| as $PO_4^{3-}$ | — | 12.06 ± 0.012 |
| Carbon dioxide | 3.48 ± 0.022 | — |
| as $CO_3^{2-}$ | — | 1.58 ± 0.010 |
| Citric acid | 0.363 ± 0.004 | — |
| as $Cit^{3-}$ | — | 0.135 ± 0.0006 |
| Chloride | 0.077 ± 0.004 | 0.022 ± 0.001 |
| Fluoride | 0.072 ± 0.003 | 0.037 ± 0.002 |
| Total | | 13.83 |
| (mEq cations)/(mEq anions) | | 1.012 |
| (mM Ca)/(mM P) | | 1.656 |

[a] From Armstrong and Singer.[15]

whatever may be the age of the tissue, bones and teeth contain calcium carbonate which is separated from the phosphate by the organic material present in these tissues. This organic barrier prevents the reaction between the hydrogen bonds and calcium combined with $CO_3^{2-}$ which occurs when the sample is heated to liberate the pyrophosphate groups.

There remains the possibility that there is a change in the chemical nature of the mineral of bone *in vivo*, as it matures during the mineralization of osteons. The terms *labile* and *stable* bone mineral have been in use for some time, but there has not been a chemical definition of these terms. There is, however, particularly in the work of Dallemagne,[20] a suggestion that as bone mineral matures it may change in chemical composition and hence in re-

activity. The hypothesis is here suggested that the highly reactive form of the bone mineral, subject to rapid ion exchange at the surfaces of the crystals, may be regarded as a calcium-deficient apatite (with hydrogen bonding accounting for the crystal structure) and that as this mineral matures with aging it is transformed into the more stable hydroxyapatite.

### 2.5. Carbonate of Bone

Although carbonate is universally present as a constituent of the mineral of normal bones and teeth, its position in the mineral is still in debate. A carbonate-containing apatite, dahllite, exists in nature and is described as carbonate hydroxyapatite. Similarly, francolite has been characterized as carbonate fluorapatite. These apatites have attracted attention because of possible analogies between their composition and crystal structure and that of the mineral of bones and teeth. It was formerly believed that $CO_3^{2-}$ could substitute for $2F^-$ or $2OH^-$, thus forming a true carbonate apatite, but this view is no longer held.

Instead, there is now the proposal, advanced and strongly supported by McConnell, that $CO_3^{2-}$ substitutes isomorphically for $PO_4^{3-}$ in the minerals francolite and dahllite. McConnell concludes that dental enamel is composed of a single mineral, carbonate hydroxyapatite (dahllite), and states that there is no valid reason for supposing that the crystal structure or crystal chemistry of bone is significantly different.[21] Against this is the view that $CO_3^{2-}$ is present in the mineral of bones and teeth only in surface positions, where it substitutes for $PO_4^{3-}$. The evidence for this hypothesis is summarized by Neuman and Neuman.[6]

Both views provide for the substitution of $CO_3^{2-}$ for $PO_4^{3-}$. Herman and Dallemagne[22] have recently studied carbonate hydroxyapatite and calcified tissues by means of infrared spectrophotometry. They find that francolite (staffelite) contains $CO_3^{2-}$ groups substituting for $PO_4^{3-}$ in the lattice, with a random distribution in the bulk of the crystal. The results with dahllite (carbonate hydroxyapatite) were different. In this mineral, a part of the $CO_3^{2-}$ substitutes for $PO_4^{3-}$, but only at the surface, while another part combines with calcium to form an adventitious calcite. Engström[2] believes the carbonate is either surface bound to the crystallites or occurs as a separate amorphous phase. If the carbonate is amorphous, this accounts for its failure to give diffraction lines. The exact state of the carbonate in the mineral of bones and teeth thus remains in doubt.

There has been little evidence to show that random substitution of $CO_3^{2-}$ for $PO_4^{3-}$, either within the apatite crystal or on the surface, is of physiological importance. This question would assume more significance if it could be shown that the presence of $CO_3^{2-}$ is essential for the formation of a mineral

comparable to that of bone. Sobel *et al.*[23] have shown that the carbonate content of the bone mineral depends upon the Ca/P ratio of the serum, as influenced by dietary Ca/P ratios and levels. The skeleton participates in buffering the carbonate (bicarbonate) of the blood and tissue fluids. Whether or not this buffering action depends upon the position of the carbonate ion in the bone mineral has not been shown.

## 2.6. Citrate of Bone

It has been known, since the report of Dickens,[24] that 90% or more of the citric acid of the body is in the skeleton and that as much as 1% of the fresh weight of bone may be accounted for as citrate.

The occurrence of citrate in the mineral portion of the skeleton forms one of the strongest arguments for the belief that such substances are held on the surfaces of apatite crystals, since the size of the citrate molecule clearly excludes it from the apatite structure itself. Whether citrate is present in bone as an ion or whether it is combined in a complex form with calcium is still not clear. That portions of the citrate may be dissolved from powdered bone only with difficulty, while other portions are readily soluble, suggests either that more than one form of citrate exists in bone or that portions of the citrate are held on entrapped surfaces.

Present interest in citrate centers upon its fundamental significance in bone. It is probable that at least part of the citrate in the circulating blood originates in the skeleton and is metabolized in the kidney. Moreover, the increase in the citrate in bone and in the blood, as induced by either the parathyroid hormone or vitamin D or by both acting together, offers presumptive evidence that citrate plays a role in the solubilization of the bone mineral and the transfer of calcium from bone to blood.[6]

## 2.7. The Mineral of Teeth

### 2.7.1. *Dentin*

The mineral of dentin, the composition of dentin matrix, and its mineralization have been studied intensively.[25] Except for the difference that dentin is acellular, its structure and composition are comparable to bone. The mineral phase, which occupies about one-fourth of the volume and four-fifths of the weight of dentin, consists of a crystalline mineral, with an apatite lattice similar to or identical with that of bone. The matrix, like that of bone, consists mainly of collagen, with cross-banding at intervals of approximately 640 A. There are also small amounts of mucopolysaccharides, similar in composition to those of cartilage. Freshly decalcified human dentin has been shown to induce formation of apatite mineral from solutions in the same manner as does

decalcified bone. While dentin is not bone, its properties are so similar to those of bone that no gross error will be committed by considering the behavior of the two tissues as identical or nearly so—the same may be said for cementum.

### 2.7.2. *Enamel*

Enamel differs fundamentally from dentin and from bone in that it is of epithelial origin and is unique to the evolution of the dental organs in higher

FIG. 5.4. Electronmicrograph of crystallites in adolescent teeth at the free surface of the enamel, close to the enamel-cement boundary. (Magnification × 240,000.) (From Rönnholm, *J. Ultrastruct. Res.*[27])

animals.[26] The morphology of enamel is treated in Chapter 2, and a comparison of its development, environment, and components is included in Table 2.1. Even with the differences between enamel and the other hard tissues of the body, the chemical composition and crystal structure of the mineral of enamel closely resemble those of bone and of dentin.

Human enamel contains the largest apatite crystals of the body (Fig. 5.4). These crystals are preferentially oriented, roughly in the long axis of the

enamel rods. The fibrous protein of the enamel matrix, existing as a three-dimensional reticulum throughout the tissue, is a eukeratin rather than a collagen and is present in extremely low amounts compared to the organic portion of bone. The total organic content of human enamel, on a weight basis, has a mean value of 0.6%.[28] Mineralization occurs slowly but continues for a long time after cessation of organic matrix formation: this results in a much larger apatite crystal size than that seen in mesenchymal hard tissues. After the enamel has reached a density of 1.5, further calcification occurs by replacement of water by mineral; the density of calcification, and consequently the hardness of the tissue, increases until the enamel resembles an almost pure apatite crystal. The mineral of enamel, like that of bone, contains carbonate. McConnell[21] concludes that enamel is composed of a single mineral, carbonate hydroxyapatite (dahllite), and that in this respect it is comparable to the mineral of bone.

## 3. Dynamics of the Bone Mineral

The preceding section has dealt with the crystal structure and chemical composition of the mineral of bones and teeth. The contributions of tracer methods to these topics have been few in number and indecisive. With respect to the dynamics of the mineral, however, the opposite is true; a large proportion of our present knowledge in this area has been derived from the use of radioisotopes. The topics studied have included nucleation, crystal formation and growth, solubility, reactivity, ion transfer and exchange, and surface chemistry.

### 3.1. Solubility Relationships

That the bone salt is difficultly soluble in the body fluids is of far-reaching physiological significance. This property not only determines the deposition of the bone mineral but also preserves the structure and rigidity of the bones. In addition, it is of first importance in maintaining the balance of minerals between the skeleton and the fluids of the body.

Numerous attempts have been made to define the solubility of the bone salt in simple physicochemical terms—i.e., by determination of a solubility product constant. These have not been successful for several reasons. First, the concept of a solubility product constant is meaningless without reference to a known and homogeneous solid phase. The bone mineral does not meet this requirement. Second, equilibrium between the solid and liquid phases of hydroxyapatite is attained *in vitro* only very slowly and with great difficulty. Moreover, the results of equilibration may differ depending on whether the state of saturation is approached from supersaturation or from undersaturation.

A part of this difficulty stems from the ability of the solid phase of the mineral to adsorb ions from solution. Third, the solid phase may undergo changes in internal composition during equilibration. Fourth, the relations between the bone mineral and the fluids bathing it are influenced by certain biological factors. Fifth, new concepts presently being introduced concerning the state of phosphate, and of its constituent ions, in the fluids of the body may influence interpretation of data obtainable by isotopic and other methods.

### 3.2. Nucleation of Bone Mineral and Crystal Growth

The nucleation of crystal formation, as the first stage of calcification of bone, has been studied largely by electron microscopy. However, radioactive phosphorus ($P^{32}$) has been employed in studies of phosphate binding to collagen,[29,30] and such methods give much promise for the future.

Since deposition of crystals of bone mineral occurs in and around collagen fibrils[3,31] and since these crystals initially have a characteristic arrangement in relation to the orientation of the fibrils and to their cross-banding, it is not surprising that collagen should be implicated in the nucleation or seeding of the crystals. In recent years it has been repeatedly demonstrated that reconstituted collagen (derived from sources such as fish swim bladder and rat tail tendon which are not ordinarily subject to calcification) may induce nucleation within its fibrils[31] (see Fig. 5.1). Rachitic osteoid[32,33] is also a favorable test object for the observation and study of nucleation (Fig. 5.5).

In both methods, nucleation depends on a metastable liquid phase, a minimal product of calcium and phosphate ions, and favorable local factors in the mineralizing tissue. Sobel *et al.*[32] state that, while available evidence suggests that collagen induces nucleation, the complete system is more complex and probably includes, among other factors, a sulfated mucopolysaccharide. Jackson and Randall[34] have illustrated the earliest stages of calcification by means of electron microscopy and have shown small isolated crystals between the bands of the collagen fibrils.

Following nucleation, there is growth of the crystals until they attain their full size. Crystal growth should be distinguished from precipitation, and the two terms should not be used interchangeably. Crystal formation emphasizes the internal structure of the crystal; attainment of this structure occurs by stepwise addition of ions to the nucleus in the course of its transformation into the final form and size of the crystal. Formation of bone mineral should be regarded as crystal formation in close relation to an organic matrix, rather than as simple precipitation within the matrix.

The internal structure of the crystals of bone mineral is determined by interionic forces and results in an array of ions characteristic for each particular mineral. It is not necessary to postulate a pattern or template which might

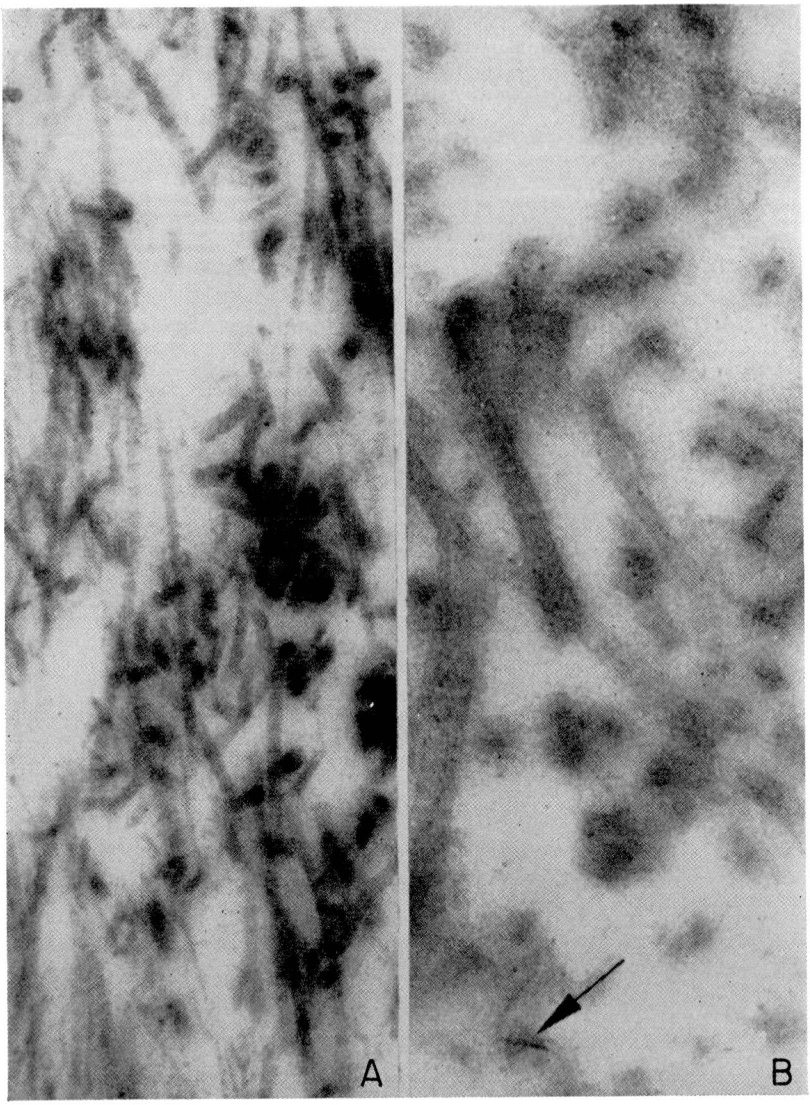


FIG. 5.5. Electronmicrographs of thin sections of rachitic osteoid. A, Control; B, Inorganic phosphate was given to the animal 2 hours before sacrifice. The small dense areas on and in the collagen fibrils are interpreted as *seeds*, the earliest visible sites of inorganic crystal formation. A larger crystal can be seen at arrow. Histological evidence of healing in B. (Magnification × 57,000.) (From Robinson and Sheldon, *Calcification in Biological Systems*.[33])

pack each ion in its special location during the nucleation and subsequent growth of a crystal. Orientation of the crystals, with their long dimensions in the axis of the collagen fibrils, does, however, suggest a mechanism for the packing of the crystals. Such orientation is especially prominent in native bone but is not demonstrable when reconstituted collagen is calcified *in vitro*. Neuman and Neuman[6] have suggested that the orientation of hydroxyapatite crystals can be influenced by *epitaxy*, which is defined as crystalline intergrowth or oriented overgrowth, and they state that the weight of evidence is that collagen, itself a crystalline substance, can produce such an epitactic growth and orientation of the crystals.

## 4. Surface Chemistry

The above considerations apply chiefly to the interior of the crystals of bone mineral. The situation on the surfaces of the crystals is quite different. $Ca^{2+}$, $PO_4^{3-}$, $CO_3^{2-}$, and $OH^-$ ions are exposed to the hydration shell and to the surrounding body fluids. Ion exchange takes place readily, especially in the more accessible and more reactive portions of the bone—viz., the new bone in the partially mineralized osteons. Here the surfaces of the crystals are in contact with an aqueous medium in which they have a finite solubility. The crystals are subject to constant ion exchange, as well as to other surface phenomena, and must be considered as participants in a very dynamic system with a labile structure. The surfaces are in equilibrium with their immediate environment; any attempt to describe them in static terms will lead to false impressions.[6]

There are recurring suggestions that the blood plasma and even the bulk of extravascular fluids are not in direct equilibrium with the crystals of the bone mineral. Instead, there is postulated a gradient, or even a physical barrier, which maintains a gradient between the pH at the surface of the crystals and that in the circulating fluids;[35] the nature of such a hypothetical barrier has not been elucidated. Nordin[36] postulates a hydrogen ion concentration corresponding to pH 6.8 at the surfaces of the crystals. Terepka *et al.*[35] state that the pH in the immediate environment of the crystals must be lower than pH 6.8 and believe that the gradient between this and the $H^+$ concentration in the circulating fluids in the body is maintained by the production of acid by the cells of bone. Borle *et al.*[37] also support the concept that the concentrations of calcium and phosphate found in the extracellular fluid are the result of local pH changes at the bone mineral surface determined by lactate formation.

We are prepared to accept the postulate that the production of acid in bone, and perhaps also the production of the citrate ion, plays a part in the solubilization of the mineral, and hence its transfer from bone to blood. We do not accept

the idea that there is a pool of fluid, at pH 6.8 or below, surrounding the crystals in process of dissolution. In fact, we regard the concept of pH itself at the crystal surfaces as one of doubtful validity owing to the small amounts of fluid bathing these surfaces. A reaction, occurring in a stepwise fashion rather than in the presence of a pool of excess acid, might be represented as follows:

$$Ca_3(PO_4)_2 + 2H^+ \rightarrow 3Ca^{2+} + 2HPO_4^{2-} \qquad (5.1)$$

It may be assumed that the conditions at the surfaces of the crystals in reactive or labile bone, such as that localized in the new and incompletely mineralized osteons, are different from those in the more stable and completely mineralized bone of the mature osteons and lamellar bone.

The special conditions at the surfaces of crystals include the effects of several forces, not mutually exclusive. The classic concept of adsorption phenomena, in which the bonding is primarily the result of van der Waals forces and in which the adsorbed material may be present in multilayers, refers primarily to the adsorption of gases by solids; ions in solution are not known to undergo classic adsorption. Chemisorption refers to electrovalent or covalent bonding, in which the chemisorbed material is usually limited to a surface monolayer. Ions in solution are frequently chemisorbed. A good example of this phenomenon is the adsorption of calcium and phosphate ions from serum shaken with lead phosphate; hydroxyapatite is also an excellent chemisorber. Many attempts have been made to explain—in terms of chemisorption—the uptake or exchange of ions by the bone mineral in contact with the fluids of the body. Currently, however, the emphasis is on exchange and transfer of ions between the body fluids and crystal surfaces.[6] This concept has been subjected to careful quantitative studies which have led to the elucidation of stoichiometric relationships and to the conclusion that the dynamics of the behavior of ions at the crystal surfaces are best expressed in terms of ion exchange without, however, eliminating the possibility of chemisorption occurring concurrently.

## References


1. F. C. McLean and M. R. Urist, *Bone: An Introduction to the Physiology of Skeletal Tissue*, 2nd ed., University of Chicago Press, Chicago, 1961.
2. A. Engström, Ultrastructure of Bone Mineral, in *Bone as a Tissue*, pp. 251-261, McGraw-Hill Book Co., Inc., New York, 1960.
3. R. A. Robinson, Electron Microscopy of Bone, in *Radioisotopes and Bone*, pp. 467-484, Blackwell Scientific Publ., Oxford, 1962.
4. Z. Molnar, Additional Observations on Bone Crystal Dimensions, *Clin. Orthopaed.* **17**: 38-42 (1960).
5. M. J. Glimcher, Specificity of the Molecular Structure of Organic Matrices in Mineralization, in *Calcification in Biological Systems*, pp. 421-487, Am. Assoc. Advancement Sci., Washington, 1960.

6. W. F. Neuman and M. W. Neuman, *The Chemical Dynamics of Bone Mineral*, The University of Chicago Press, Chicago, 1958.
7. A. Engström and J. B. Finean, *Biological Ultrastructure*, Academic Press Inc., New York, 1958.
8. S. B. Hendricks and W. L. Hill, The Inorganic Constitution of Bone, *Science*, **96**: 255-257 (1942).
9. M. L. Dallemagne and H. Brasseur, La Diffraction des Rayons X par la Substance Minérale Osseuse, *Bull. Soc. Roy. Sci. Liège*, **11**: 451-462 and 488-495 (1942).
10. A. S. Posner, C. Fabry, and M. J. Dallemagne, Defect Apatite Series in Synthetic and Natural Calcium Phosphates: the Concept of Pseudoapatites, *Biochim. Biophys. Acta*, **15**: 304-305 (1954).
11. A. S. Posner and A. Perloff, Apatites Deficient in Divalent Cations, *J. Res. Natl. Bur. Std.*, **58**: 279-286 (1957).
12. A. S. Posner, J. M. Stutman, and E. R. Lippincott, Hydrogen-bonding in Calcium-deficient Hydroxyapatites, *Nature (London)*, **188**: 486-487 (1960).
13. L. Winand, *Etude Physico-chimique des Phosphates de Calcium de Structure Apatitique*, Thesis, University of Liège, 1960.
14. H. Herman, P. François, and C. Fabry, Le Composé Minéral Fondamental des Tissus Calcifiés. I.—Présence de Groupements Acides dans le Réseau Apatitique des Phosphates de Calcium Synthétiques, *Bull. Soc. Chim. Biol.*, **43**: 629-642 (1961).
15. W. D. Armstrong and L. Singer (unpublished).
16. P. François and H. Herman, Le Composé Minéral Fondamental des Tissus Calcifiés. II.—Les Sels Osseux Contiennent un Phosphate de Calcium Différent de l'Hydroxylapatite, *Bull. Soc. Chim. Biol.*, **43**: 643-649 (1961).
17. P. François, Étude de la Variation de Composition de l'Os du Rat avec l'Âge, *J. Physiol. (Paris)*, **53**: 343-344 (1961).
18. H. Herman and M. J. Dallemagne, The Main Mineral Constituent of Bone and Teeth, *Arch. Oral Biol.*, **5**: 137-144 (1961).
19. H. Herman, A Propos du Taux de Pyrophosphate de l'Os et des Tissus Dentaires Soumis au Traitement Thermique, *Arch. Intern. Physiol. Biochim.*, **69**: 748-749 (1961).
20. M. J. Dallemagne, Les Sels Osseux. Etat Actuel de la Question. *Bull. Acad. Roy. Med. Belg.*, VII[e] Serie, **1**: 749-808 (1961).
21. D. McConnell, Recent Advances in the Investigation of the Crystal Chemistry of Dental Enamel, *Arch. Oral Biol.*, **3**: 28-34 (1960).
22. H. Herman and M. J. Dallemagne, Carbonato-hydroxyapatites and Teeth and Bone Carbonate as Studied by Infra-red Spectrometry, *Arch. Oral Biol.* (in press).
23. A. E. Sobel, M. Rockenmacher, and B. Kramer, Carbonate Content of Bone in Relation to the Composition of Blood and Diet, *J. Biol. Chem.*, **158**: 475-489 (1945).
24. F. Dickens, The Citric Acid Content of Animal Tissues, with Reference to Its Occurrence in Bone and Tumour, *Biochem. J.*, **35**: 1011-1023 (1941).
25. C. C. Solomons, J. T. Irving, and W. F. Neuman, Calcification of the Dentin Matrix, in *Calcification in Biological Systems*, pp. 203-216, Am. Assoc. Advancement Sci., Washington, 1960.
26. R. M. Frank, R. F. Sognnaes, and R. Kern, Calcification of Dental Tissues with Special Reference to Enamel Ultrastructure, in *Calcification in Biological Systems*, pp. 163-202, Am. Assoc. Advancement Sci., Washington, 1960.
27. E. Rönnholm, The Amelogenesis of Human Teeth as Revealed by Electron Microscopy. II. The Development of the Enamel Crystallites. *J. Ultrastruct. Res.*, **6**: 249-303 (1962).

28. M. V. Stack, The Chemical Nature of the Organic Matrix of Bone, Dentin, and Enamel, *Ann. N. Y. Acad. Sci.*, **60**: 585-595 (1955).

29. M. J. Glimcher and S. M. Krane, Studies of the Interactions of Collagen and Phosphate. I. The Nature of Inorganic Orthophosphate Binding, in *Radioisotopes and Bone*, pp. 393-418, Blackwell Scientific Publ., Oxford, 1962.

30. S. M. Krane and M. J. Glimcher, Studies of the Interactions of Collagen and Phosphate. II. Nucleotidase Activity and Binding of Nucleotide Phosphorus, in *Radioisotopes and Bone*, pp. 419-442, Blackwell Scientific Publ., Oxford, 1962.

31. M. J. Glimcher, Molecular Biology of Mineralized Tissues with Particular Reference to Bone, *Rev. Modern Phys.*, **31**: 359-393 (1959).

32. A. E. Sobel, M. Burger, and S. Nobel, Mechanisms of Nuclei Formation in Mineralizing Tissues, *Clin. Orthopaed.*, **17**: 103-123 (1960).

33. R. A. Robinson and H. Sheldon, Crystal-collagen Relationships in Healing Rickets, in *Calcification in Biological Systems*, pp. 261-279, Am. Assoc. Advancement Sci., Washington, 1960.

34. S. F. Jackson and J. T. Randall, Fibrogenesis and the Formation of Matrix in Developing Bone, in *Ciba Foundation Symposium on Bone Structure and Metabolism*, pp. 47-64, Little, Brown & Co., Boston, 1956.

35. A. R. Terepka, C. M. Dowse, and W. F. Neuman, *Unifying Concepts of Parathyroid Hormone Action*, USAEC Report UR-577, pp. 1-28, Univ. Rochester Atomic Energy Project, September 1960.

36. B. E. C. Nordin, The Solubility of Powdered Bone, *J. Biol. Chem.*, **227**: 551-564 (1957).

37. A. B. Borle, N. Nichols, and G. Nichols, Jr., Metabolic Studies of Bone *in vitro*. I. Normal bone, *J. Biol. Chem.*, **235**: 1206-1210 (1960).

of bone were studied autoradiographically and by means of microradiograms. In observations made 4 to 6 weeks after injection of $Na^{22}$, there was concentration of the radioisotope in osteons in which bone mineral was being deposited at the time of injection. This is evidence for deposition of new sodium during initial mineralization. After immersion in a solution of $Na^{22}Cl$, *in vitro*, compact bone was found to be uniformly labeled (Fig. 6.1).

## 3. Magnesium in Bone

Although there are numerous papers concerning the distribution of magnesium in bone, there is little information on homeostatic regulation of the magnesium content of the fluids of the body. Recent investigations utilizing $Mg^{28}$ with a half-life of 21.4 hours have added to an understanding of the kinetics of the behavior of this element in the organism.

Silver *et al.*,[8] in a study of the turnover of $Mg^{28}$ in man, conclude that there is a relatively large pool—made up largely of the magnesium of bone—with a slow turnover rate. MacIntyre *et al.*,[9] who were also using $Mg^{28}$, believe that they have refuted the concept of the skeleton of man as a reservoir of available magnesium. These authors suggest that in adult man bone magnesium may be largely unavailable, which is consistent with the very slow turnover rate. Breibart *et al.*[10] have found that the exchange of $Mg^{28}$ in rat cortical bone occurs much more rapidly in young animals than in old ones. Glaser and Gibbs[11] have studied the localization of radiomagnesium in puppies by utilizing autoradiography of bone. Magnesium-28 produced the highest densities in the epiphyseal line (in agreement with the results of radioassay, which demonstrated a higher concentration of $Mg^{28}$ in the epiphyseal line than in the diaphyses or the epiphyses).

Taylor,[12] in a study of the stable magnesium of bone, has described fractionation experiments in which powdered bone was extracted with dilute acids and also with ethylenediamine tetraacetic acid (EDTA). He concludes that at least 70% of the total bone magnesium is located on the surfaces of the bone crystals, either as $Mg^{2+}$ or $MgOH^{+}$ ions. This is consistent with the earlier conclusion of Duckworth *et al.*[13] that a third of the skeletal magnesium can be mobilized in severe magnesium deficiency, but is not in agreement with the more recent view, derived from isotope studies with $Mg^{28}$, of the relative unavailability of skeletal magnesium.

## 4. Zinc in Bone

The role of zinc in animal nutrition and metabolism has been reviewed in detail by Asling and Hurley,[14] who paid special attention to skeletal tissue. Vallee[15] has summarized the present state of the subject of the association of

zinc with a number of enzymes, both as an integral part of the enzyme molecule (metalloenzyme) and as an activating, loosely bound cofactor (metal-enzyme complexes). Of the nine zinc enzymes listed by Vallee as having been identified and characterized, two are of possible interest in relation to the mineralization of bone—carbonic anhydrase and alkaline phosphatase. Zinc, as a trace element, is essential to growth and nutrition in animals, including man. The nature of its activity, especially in the growth and remodeling of bone, is receiving further attention with the aid of radioactive zinc, $Zn^{65}$.

Haumont[16] has studied the remodeling of compact bone and the endochondral growth of bone in various species by using the zinc-specific stain dithizone. He has shown that zinc is deposited in haversian systems in the process of formation, in a strongly positive line at the junction of the uncalcified preosseous tissue with the newly calcified bone, and concludes that zinc is deposited in this location at exactly the time that calcification begins. According to his interpretation, zinc is not incorporated within the bone mineral but, rather, is either acting in an enzyme system or acting directly, as a metallic ion, in crystal seeding or nucleation. Once deposited, zinc persists in calcified bone, although the nature of its combination in mature bone is still uncertain.

Haumont also studied the distribution of $Zn^{65}$ in bone after its administration to experimental animals. Precise localization of this isotope has been established by comparing microradiograms with autoradiograms of the same sections, as shown in Fig. 6.2. The radiozinc is found in the layers of preosseous tissue about to be calcified, i.e., the border of the preosseous tissue as stained by histochemical methods. Radiozinc is not detected in the bone tissue that was present at the time of injection, it does not produce any diffuse reaction in the autoradiographic emulsion, and it is not exchanged with the constituents of bone once calcified. By extension of the autoradiographic study to the growth apparatus of long bones, the most intense activity has been found at the level of the calcification front; lesser activity was observed in areas of periosteal accretion. It is suggested that zinc may affect some enzyme system in the initial stage of calcification although such an enzymatic mechanism has not been identified.

Zinc has also been studied in relation to its ability to inhibit mineralization of rachitic cartilage *in vitro*.[17] Of the elements tested, including manganese, cobalt, cadmium, and chromium, in addition to zinc, the most potent inhibitor of mineralization was zinc, and its presence in small amounts enhanced the inhibitory effect of the other ions on mineralization. That zinc may both promote and inhibit mineralization of osseous tissue suggests that the local concentration is the factor determining the action in a particular instance.

Prasad *et al.*[18] have reported on 17 zinc-deficient patients, from villages

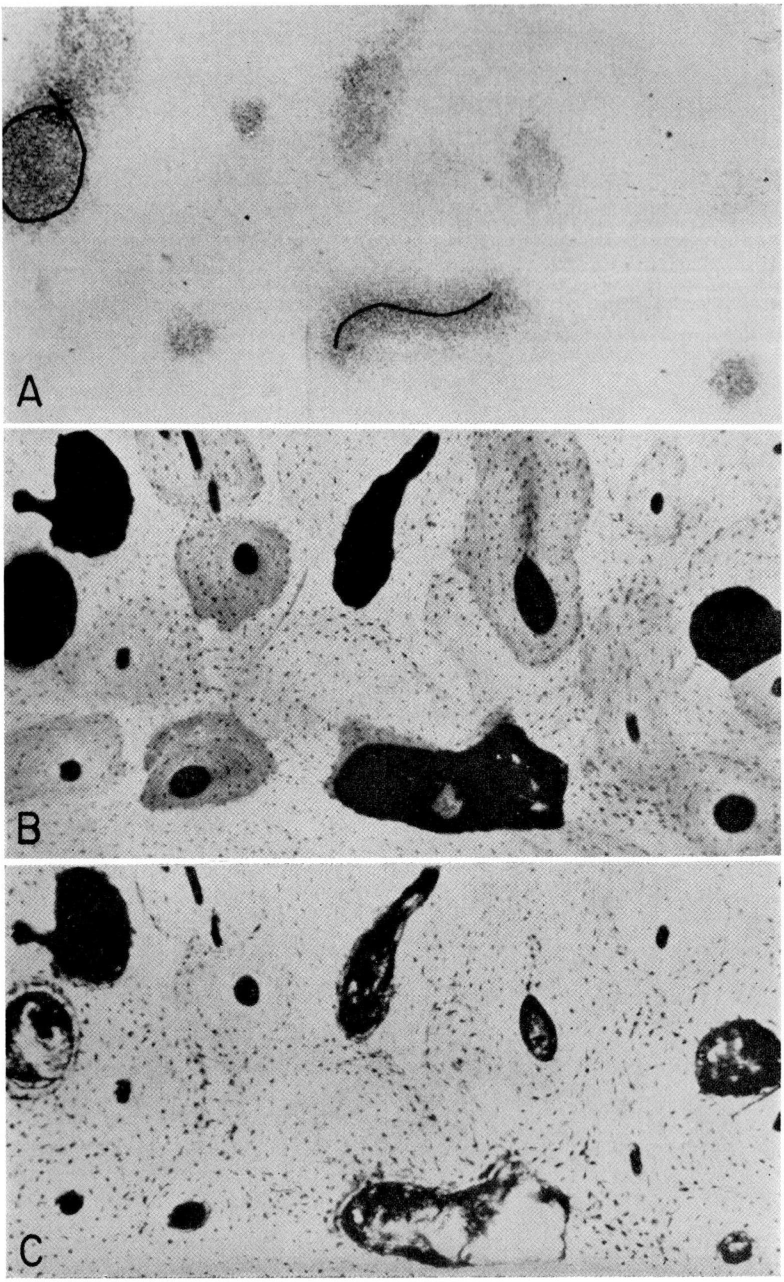


FIG. 6.2. Correlation between autoradiograph (A), microradiograph (B), and histologic section (C) of a transverse section of a monkey humerus 5 hours after injection of $Zn^{65}$. The radioactivity appears in the border in the preosseous tissue undergoing calcification. (Magnification × 85.) (From Haumont, *Le Zinc dans le Tissu Osseux.*[16])

near Cairo, Egypt, who exhibited the syndrome of dwarfism, hypogonadism, hepatosplenomegaly, and iron-deficiency anemia. These patients exhibited low plasma zinc, high plasma zinc turnover, small 24-hour exchangeable zinc pool, decreased excretion of zinc in stool and urine, and low content of zinc in the hair. Serum alkaline phosphatase was low. Growth retardation, hypogonadism, and partial adrenal hypofunction were attributed to hypofunction of the anterior pituitary, conditioned by zinc deficiency. Anemia was of the iron-deficiency type. The bones were not studied, except that X-rays showed that the epiphyses were not closed.

## 5. Fluoride in Bone

The prototype of the apatites in naturally occurring minerals is *fluorapatite*, in which the fluoride ion, $F^-$, is present instead of the $OH^-$ of hydroxyapatite. Owing to the presence of fluoride in variable amounts in drinking water, the mineral of bone has a small admixture of fluorapatite. This situation has received special attention in relation to the teeth, which appear to be the body organs most sensitive in terms of response to small increments in fluoride

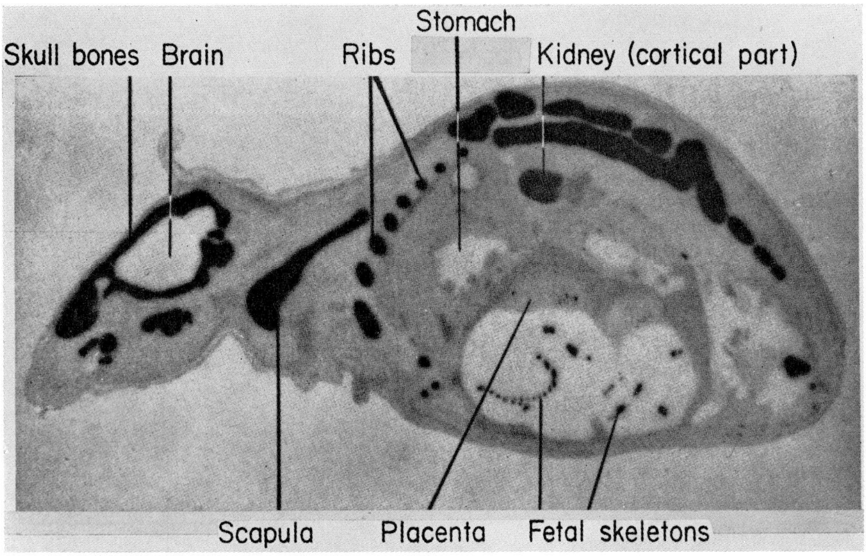


FIG. 6.3. Autoradiograph of a pregnant mouse 30 minutes after intravenous injection of $F^{18}$. Most of the radioactivity is taken up in the skeleton; some has passed the placenta and appears in the fetal bones (full term fetuses with beginning mineralization). The fetal soft tissues appear totally devoid of radiofluorine. (Magnification × 1.1.) (From Ericsson and Ullberg, *Acta Odontol. Scand.*[20])

ingestion; uptake of fluoride by the teeth is now recognized as important in the prevention of caries.

The usefulness of radiofluorine, $F^{18}$, in the study of bones and teeth has been limited by its short half-life (109.7 minutes).[19] By the use of methods for rapid isolation and purification of $F^{18}$ combined with a rapid method for sectioning of whole frozen animals, Ericsson and Ullberg[20] have been able to obtain excellent autoradiograms that demonstrate the isotope in bones and teeth. The bones take up radiofluorine within a few minutes following intravenous injection. Uptake is mainly in the zones of mineralization; uncalcified cartilage shows no affinity for $F^{18}$ (Fig. 6.3). The substitution of $F^{18}$ in the apatite crystal structure to form fluorapatite aids in the regulation of the plasma fluoride concentration, with varying intake of the fluoride ion. Perkinson *et al.*[21] assume that the observed deposition of $F^{18}$ for the most part does not represent incorporation into newly forming bone crystals but can be accounted for by ion exchange, probably with $OH^-$ groups of the apatite structure. This concept is consistent with the role of the skeleton in homeostasis as described by Singer and Armstrong.[22]

Under the usual conditions of low intake in drinking water, the skeletal content of fluoride increases with age; pathological fluorosis of the skeleton is rare and is associated only with continued high intake of the element.[23]

## 6. Other Trace Elements and Bone

A distinction must be made between *trace elements* and radioelements given in tracer amounts. A trace element, of which more than 20 have been identified, is defined as one which is constantly present in the animal organism, usually in low but variable concentrations. Of these elements, Fe, Cu, Mn, Zn, I, and Co have been shown to be nutritionally essential for the higher forms of animal life, although there is some justification for including Mo, F, Ba, and Sr within this group.[24] Of the ten elements listed, four (including Mn, Zn, Cu, and I) are required for normal skeletal development in a variety of birds and mammals,[14] while a fifth, Mo, has deleterious effects on the skeleton when it is present in the diet in excess.

Of these ten elements, two (Zn and F) have been considered above in relation to the specificity of their functions in bone and more particularly in relation to the use of their isotopes in the study of bone. Further consideration will be given in this section to the role of the additional trace elements in skeletal physiology.

As to radioelements given in tracer quantities, Durbin[25] has reviewed the available data for the biological behavior in rats of high specific activity radioisotopes of 70 elements. Tracer quantities were administered in single

intramuscular injections, and the data obtained by radioassay of tissues and excreta were categorized to bring out the metabolic characteristics of the elements within a chemical family. Most of the elements studied were found in bone as well as in other tissues, and the data obtained with elements of special interest to bone have been discussed in Chapter 3.

## 6.1. Manganese

In contrast to the specific localization and probable function of zinc in bone as described above, the role of manganese in skeletal metabolism remains obscure. All observers are agreed that the highest concentrations of this element are found in bone. The average of figures in the literature recorded for the rabbit is 3.5 mg of Mn per kilogram of wet bone[26] as compared with 1.9 mg per kilogram of fresh liver. Following administration of $Mn^{54}$ as $MnCl_2$, significant amounts of Mn were found in spleen, pancreas, kidney, and skeleton,[25] but it was not determined whether the skeletal Mn was deposited as colloidal particles in the marrow or was present in the bone tissue itself in a manner similar to other divalent elements. Fore and Morton[27] reported that the manganese in ox bone is of the order of 0.27 to 0.51 mg per kilogram, corresponding to about 1.4 mg per kilogram of ash. These authors concluded that, while most of the manganese was deposited with the inorganic salts, a small proportion was definitely associated with the organic matrix.

Cotzias,[28] in a comprehensive review, concludes that the pathway of manganese through the body is specific and seems to pass through the intracellular organelles. He suggests that manganese metabolism is homeostatically controlled, as has also been indicated by the dependence of manganese turnover upon the total body burden of the metal. It appears that the requirement for trace amounts of this element will have to be accounted for on a microchemical level. The actions of manganese *in vitro*, including enzyme activation, transport systems, and preservation of mitochondria, can be duplicated with other metals. Identification of the actions of manganese essential to higher organisms has not yet been possible.

## 6.2. Copper

While the presence of copper has been demonstrated in the skeleton, its role in bone is still obscure. Copper has been found in enamel, and it occurs normally in the brown teeth of beavers.[29] The best known actions of copper are related to iron absorption from the gastrointestinal tract and to hemoglobin formation. Infants as well as rats receiving only a milk diet develop an iron-deficiency anemia. In infants, the deficiency is a microcytic, normochromic anemia caused by failure of the bone marrow to synthesize an adequate quantity of hemoglobin.

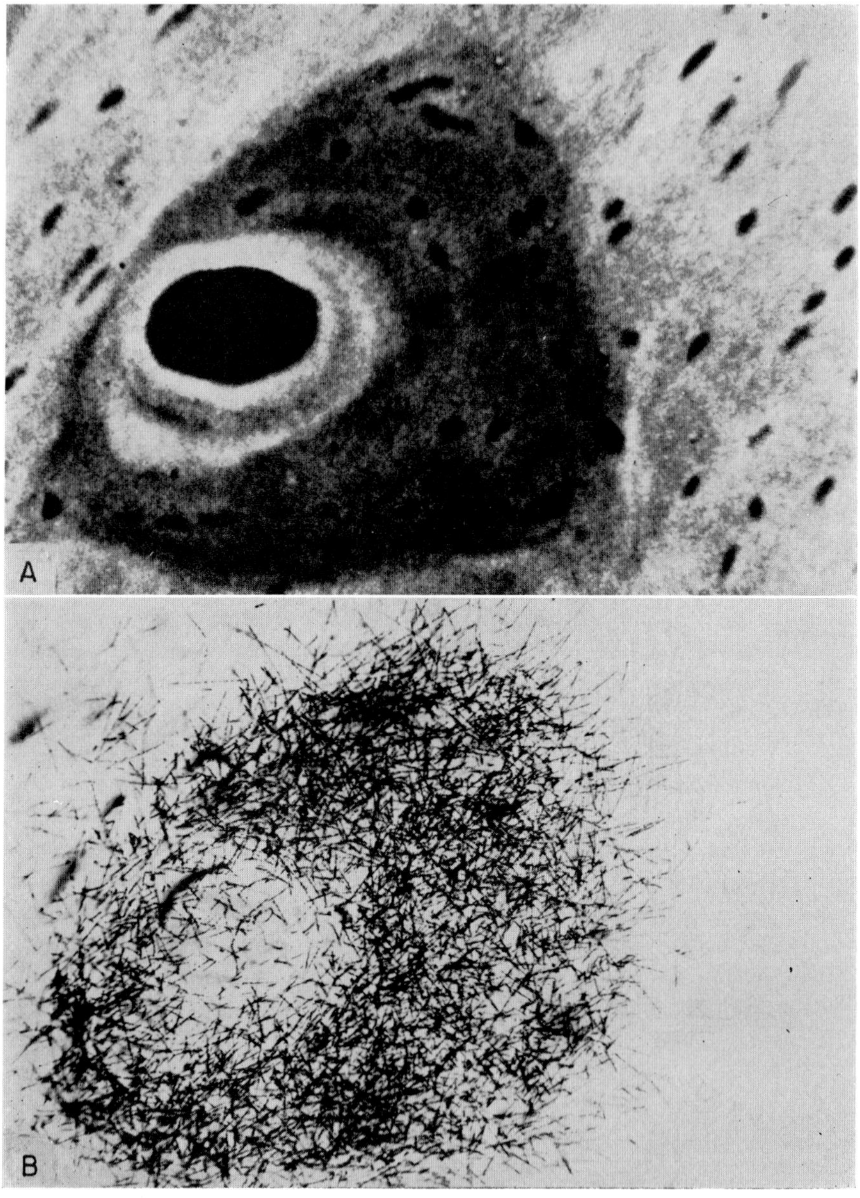


FIG. 6.4. Microradiograph (A) and autoradiograph (B) of a single haversian system. This is an example of a completely radium-labeled system. (Magnification × 430.) (From Rowland and Marshall, *Radiation Res.*[31])

Copper is required for normal skeletal development in sheep. In continued copper deficiency, rats and pigs develop a rickets-like syndrome in addition to neurological disturbances. The effects of copper deficiency in herbivores are aggravated by a concurrent excess of molybdenum.[14] The concentrations of copper in subcellular fractions of rat liver, in contrast with those of manganese, are very low in the particulate fractions, especially the mitochondria and the microsomes, and very high in the soluble fraction.[30] As in the case of manganese, it seems likely that copper exerts its effects at the microchemical or the intracellular level.

## 7. Radium and Plutonium in Bone

Observations made on the distribution of radium in bone, both in experimental animals and in persons exposed to accidental radium poisoning, have given information of much interest, especially with respect to the diffuse component of radioisotope deposition which resembles that of radiocalcium. The activity in the diffuse distribution in bone samples from individuals who have carried radium 20 years or more has been found to be relatively uniform in cortical bone (Fig. 6.4). The average magnitude of this activity is about one-half of the total body burden divided by the weight of the bones. The specific activities in the hotspots are, on the average, 90 times as great as in the diffuse component. The total amount carried in the hotspots is, however, approximately equal to that in the diffuse component when the entire skeleton is taken into account.[31,32]

Plutonium-239, a man-made element, has an atomic ($Z$) number of 94. Although at the time it was created for the atomic bomb, plutonium was the heaviest element known, it has since been superseded in this respect by several additional artificially formed elements. It has a half-life of 24,000 years and emits alpha, beta, and gamma radiation. When introduced into the organism, plutonium has a special affinity for bone but not for the bone mineral with which it does not associate itself. Instead plutonium deposits in the organic matrix of bone, especially in the endosteal surfaces, and has been found within the cytoplasm of osteoclasts.[33]

### References

1. D. L. Buchanan and A. Nakao, Studies on the Nature of Bone Carbonate, *Arch. Biochem. Biophys.*, **77**: 168-180 (1958).
2. G. Nichols, Jr. and N. Nichols, The Role of Bone in Sodium Metabolism, *Metab.*, **5**: 438-446 (1956).
3. I. S. Edelman, A. H. James, H. Baden, and F. D. Moore, Electrolyte Composition of Bone and the Penetration of Radiosodium and Deuterium Oxide into Dog and Human Bone, *J. Clin. Invest.*, **33**: 122-131 (1954).

4. G. C. H. Bauer, Metabolism of Bone Sodium in Rats Investigated with $Na^{22}$, *Acta Physiol. Scand.*, **31**: 334-350 (1954).
5. G. C. H. Bauer, Rate of Bone Salt Formation in a Healing Fracture Determined in Rats by Means of Radiocalcium, *Acta Orthopaed. Scand.*, **23**: 169-191 (1954).
6. G. B. Forbes, G. L. Mizner, and A. Lewis, Effect of Age on Radiosodium Exchange in Bone (Rat), *Am. J. Physiol.*, **190**: 152-156 (1957).
7. J. Vincent, Autoradiographic Studies on Bone Sodium, in *Radioisotopes and Bone*, pp. 265-275, Blackwell Scientific Publ., Oxford, 1962.
8. L. Silver, J. S. Robertson, and L. K. Dahl, Magnesium Turnover in the Human Studied with $Mg^{28}$, *J. Clin. Invest.*, **39**: 420-425 (1960).
9. I. MacIntyre, S. Hanna, C. C. Booth, and A. E. Read, Intracellular Magnesium Deficiency in Man, *Clin. Sci.*, **20**: 297-305 (1961).
10. S. Breibart, J. S. Lee, A. McCoord, and G. B. Forbes, Relation of Age to Radiomagnesium Exchange in Bone, *Proc. Soc. Exptl. Biol. Med.*, **105**: 361-363 (1960).
11. W. Glaser and W. D. Gibbs, Localization of Radiomagnesium in Puppies: Radioautographic Study of Heart and Bone, *Am. J. Physiol.*, **202**: 584-588 (1962).
12. T. G. Taylor, The Magnesium of Bone Mineral, *J. Agr. Sci.*, **52**: 207-216 (1959).
13. J. Duckworth, W. Godden, and G. M. Warnock, The Effect of Acute Magnesium Deficiency on Bone Formation in Rats, *Biochem. J.*, **34**: 97-108 (1940).
14. C. W. Asling and L. S. Hurley, The Influence of Trace Elements on the Skeleton, *Clin. Orthopaed.*, **27**: 213-264 (1963).
15. B. L. Vallee, Biochemistry, Physiology and Pathology of Zinc, *Physiol. Rev.*, **39**: 443-490 (1959).
16. S. Haumont, *Le Zinc dans le Tissu Osseux*, Arscia, Brussels, 1962.
17. E. D. Bird and W. C. Thomas, Jr., Effect of Various Metals on Mineralization *in vitro*, *Proc. Soc. Exptl. Biol. Med.*, **112**: 640-643 (1963).
18. A. S. Prasad, A. Mirale, Jr., Z. Farid, H. H. Sandstead, A. R. Schulert, and W. J. Darby, Biochemical Studies on Dwarfism, Hypogonadism, and Anemia, *Arch. Int. Med.*, **111**: 407-428 (1963).
19. C. H. Carlson, L. Singer, D. H. Service, and W. D. Armstrong, Preparation of Carrier Free Radiofluoride with a New Estimate of the Half-life of $F^{18}$, *Intern. J. Appl. Radiation Isotopes*, **4**: 210-213 (1959).
20. Y. Ericsson and S. Ullberg, Autoradiographic Investigations of the Distribution of $F^{18}$ in Mice and Rats, *Acta Odontol. Scand.*, **16**: 363-374 (1958).
21. J. D. Perkinson, Jr., I. B. Whitney, R. A. Monroe, W. E. Lotz, and C. L. Comar, Metabolism of Fluorine 18 in Domestic Animals, *Am. J. Physiol.*, **182**: 383-389 (1955).
22. L. Singer and W. D. Armstrong, Regulation of Human Plasma Fluoride Concentration, *J. Appl. Physiol.*, **15**: 508-510 (1960).
23. P. S. Chen, Jr., A. R. Terepka, and H. C. Hodge, The Pharmacology and Toxicology of the Bone Seekers, *Ann. Rev. Pharmacol.*, **1**: 369-396 (1961).
24. E. J. Underwood, *Trace Elements in Human and Animal Nutrition*, Academic Press Inc., New York, 1956.
25. P. W. Durbin, Metabolic Characteristics Within a Chemical Family, *Health Phys.*, **2**: 225-238 (1960).
26. H. Fore and R. A. Morton, Manganese in Rabbit Tissues, *Biochem. J.*, **51**: 600-603 (1952).
27. H. Fore and R. A. Morton, The Manganese in Bone, *Biochem. J.*, **51**: 598-600 (1952).
28. G. C. Cotzias, Manganese in Health and Disease, *Physiol. Rev.*, **38**: 503-532 (1958).

29. R. F. Sognnaes, Dental Aspects of the Structure and Metabolism of Mineralized Tissues, in *Mineral Metabolism*, Vol. I, pp. 676-741, Academic Press Inc., New York, 1961.
30. R. E. Thiers and B. L. Vallee, Distribution of Metals in Subcellular Fractions of Rat Liver, *J. Biol. Chem.*, **226**: 911-920 (1957).
31. R. E. Rowland and J. H. Marshall, Radium in Human Bone: The Dose in Microscopic Volumes of Bone, *Radiation Res.*, **11**: 299-313 (1959).
32. J. H. Marshall, Microscopic Metabolism of Calcium in Bone, in *Bone as a Tissue*, pp. 144-155, McGraw-Hill Book Co., Inc., New York, 1960.
33. J. S. Arnold and W. S. S. Jee, Bone Growth and Osteoclastic Activity as Indicated by Radioautographic Distribution of Plutonium, *Am. J. Anat.*, **101**: 367-418 (1957).

# Chapter 7

# *Kinetics of the Bone Mineral*

## 1. Introductory Statement

While radioisotopes have been of limited value in elucidating the structure of bone mineral, they have assumed first-rate importance in the study of the kinetics of the mineral, especially in connection with the uptake, retention, and deposition of calcium and of other bone-seeking elements. In fact, the entire foundation of the kinetics of the bone mineral now rests upon data obtained by the use of isotopic materials and methods.

Until Chiewitz and Hevesy first showed[1] that there is a rapid interchange of phosphate between bone and blood, it was generally assumed that the bone mineral, once formed, remained in a fixed state until it was removed. Moreover, there was no appreciation of the rapidity with which calcium and phosphate leave the blood plasma and are replaced with similar ions which originate mainly from the bones.

Significant progress with reference to the kinetics of the disappearance of calcium from the blood awaited the availability of $Ca^{45}$. Norris and Kisieleski[2] demonstrated that, after intravenous injection of $Ca^{45}$ in mature rats, there was a rapid fall in the specific activity of the blood and a rapid rise in that of the bones. The specific activity in the bones reached a maximum in about 100 minutes and did not fall appreciably in the next 10 days. The specific activity in the blood dropped throughout the experimental period; after 10 days, the specific activity of plasma calcium was below that of bone calcium.

Armstrong *et al.*[3] published blood disappearance curves for $Ca^{45}$ in dogs. They found that the curves could be expressed as a series of four exponentials, and similar curves for $Na^{22}$, obtained simultaneously in some of the same dogs, were described by a series of three exponentials. The data were also used to estimate the quantities of skeletal calcium and tissue sodium exchanged with the blood.

Thomas *et al.*[4,5] published similar data obtained from rabbits and also found that their curves could be described by a series of four exponentials for adult rabbits while two exponentials sufficed for young rabbits. Hansard *et al.*[6] used kinetic methods and $Ca^{45}$ to study the effects of age upon the physiological behavior of calcium in cattle, and published blood-disappearance

curves for animals of different ages. These workers did not fit exponentials to their curves, but instead they calculated the rate of disappearance of calcium from the blood and arrived at figures of 102% per minute for cattle at the age of 10 days, 72% at 6 months, and 26% at 160 months. Bronner *et al.*[7] studied the fate of intravenously injected $Ca^{45}$ in nine boys and one young adult and again described the disappearance curves by a series of four exponentials. Their equations yielded a disappearance rate for serum calcium of approximately 19% per minute, or somewhat less than the figure obtained for cattle by Hansard *et al.*[6] In this connection, it is of interest that Bauer *et al.*[8] state that one calcium ion out of four leaves the blood every minute. It may be said that by 1956 the disappearance of $Ca^{45}$ from blood, following a single intravenous injection, had been well described for dogs, rabbits, cattle, and man.

## 2. Kinetic Studies — Interpretation

While much of the above-described work was in progress, Bauer, Carlsson, and Lindquist had been working on various aspects of skeletal metabolism and had published a number of papers bearing on this subject, mainly reporting on isotopic experiments on rats. In 1955 they scored a major breakthrough, by relating the disappearance of $Ca^{45}$ from the blood to its uptake by bone by a series of equations by which it became possible to calculate the rates of accretion, resorption, and exchange reactions in the skeleton—parameters previously inaccessible to analysis.[9] The methods for studying the kinetics of calcium metabolism and of the bone mineral have since been extended and refined, in large part by the original authors but also by contributions from other workers. The present state of knowledge on the metabolism and homeostatic function of bone has been extensively studied and reviewed by Bauer *et al.*,[8] and will therefore be only briefly considered here. Studies of theory and of interpretation will be reviewed first, and then followed by a discussion of the results obtained.

The original method, as published in 1955 by Bauer and his associates,[9] depended upon (1) determination of the $Ca^{45}$ content of samples of bone and of the specific activity of the serum (plasma) calcium in rats at intervals following injection of a single dose of $Ca^{45}$ and upon (2) the following basic equation:

$$Ca^{45}_{Obs} = Ca^{45}_{E} + Ca^{45}_{A} - Ca^{45}_{R} \qquad (7.1)$$

where $Ca^{45}_{Obs}$ = total amount of $Ca^{45}$ present in a calcified tissue

$Ca^{45}_{E}$ = amount of $Ca^{45}$ present in the exchangeable fraction of the bone salt

$Ca_A^{45}$ = amount of $Ca^{45}$ incorporated into the nonexchangeable fraction of the bone salt through accretion

$Ca_R^{45}$ = amount of $Ca^{45}$ removed through resorption

In more recent publications, while adhering to the same principles, Bauer and his co-workers have made changes in the symbols and in the methods of calculation to conform with common usage; the treatment here will follow that in their 1961 publication.[8]

Their general scheme for the kinetics of skeletal calcium, as indicated by tracer studies in animals, is summarized as follows.

If the period of observation is shorter than the lifespan of the tissue examined, two types of movement of calcium appear to be detectable: (1) a rapid *exchange* between the plasma calcium and a fraction of the bone calcium and (2) a one-way movement of calcium from plasma to bone (*accretion*). When the period of observation is longer than the lifespan of the tissue examined, a third process enters the picture, i.e., loss of calcium atoms from the tissue, apparently caused by *resorption* (bones) or *attrition* (incisors of rodents). This summary is derived from Bauer's work on bone samples from animals (chiefly rats) but the same principles are applicable to study of the entire body, including that of man.

The first situation, in which only two types of movement are detectable, is illustrated in Fig. 7.1 in which the curve for $E^* + A^*$ is the sum of one curve ($E^*$) which has the same shape as the serum activity curve and another curve ($A^*$) which has the same shape as the integrated serum activity curve. Without assigning physical meaning to these terms, we may express these curves, based on experimental observations, as follows:

$$B_t^* = k_1 S_t^* + k_2 \int_0^t S^* \, (t) \, dt \tag{7.2}$$

where $B_t^*$ = the amount of radioisotope in tissue $B$ (bone) at time $t$

$S_t^*$ = the specific activity of the element (i.e., activity per unit weight of the element) in serum at time $t$

$\int_0^t S^* \, (t) \, dt$ = the integrated serum activity, i.e., the area under the specific activity curve of the element in serum between $t = 0$ and time $= t$, and $k_1$ and $k_2$ are constants

The authors arrive at the conclusion that, within acceptable limits, $k_1$ may be taken as equal to the amount of exchangeable bone calcium in the sample with the symbol $E$. Similarly, $k_2$ is taken as equal to $A$, or the rate of accretion. With these stipulations, Eq. (7.2) may be written as follows:

$$B_t^* = ES_t^* + A \int_0^t S^* \, (t) \, dt \tag{7.3}$$

This equation now contains two constants, $E$ and $A$, and two variables, $B_t^*$ and $S_t^*$. These two variables can be determined experimentally, and the two constants can thus be computed by using the data observed at two time intervals.

When longer time intervals are chosen, $k_1$ and $k_2$ are found to be time-dependent, and a correction is introduced for the resorption of bone contain-

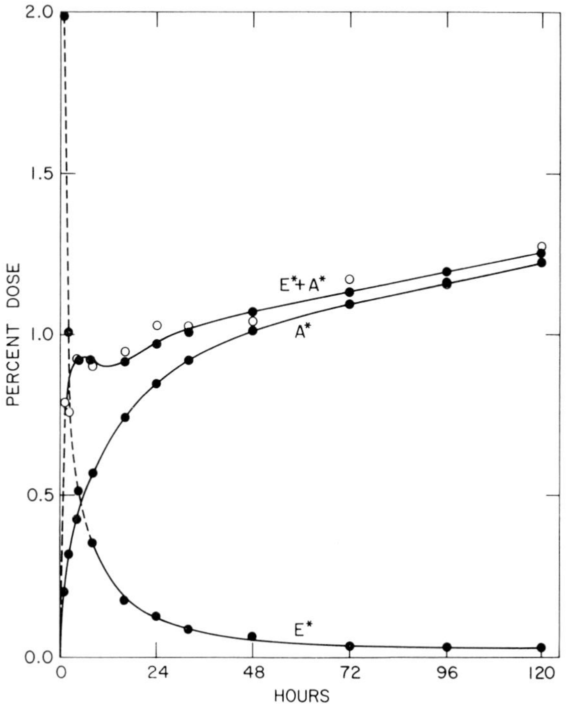


FIG. 7.1. Role of accretion and exchange reactions in the uptake of $Ca^{45}$ in the tibial shafts of young rats. (From Bauer *et al.*, *Mineral Metabolism*.[8])

ing $Ca^{45}$. The curve for the apparent loss by resorption is shown in Fig. 7.2 and is represented by the equation:

$$R_t^* = E_t^* + A_t^* - B_t^* \quad (7.4)$$

in which $R_t^*$ represents the amount of $Ca^{45}$ removed by resorption.

The above equations were originally derived from experiments on animals and required the analysis of samples of blood and of bone. For extension to the study of whole man, serial blood samples have been analyzed and $B_t^*$ of Eq. (7.4) is taken to represent the amount of isotope retained in the entire skeleton (i.e., dose administered minus excretion) at time $t$. The value of

*E* then includes both extraskeletal and intraskeletal exchangeable calcium, and corresponds to the miscible calcium pool referred to below. Bauer *et al.*,[10] from their own data and those of others, calculated the accretion rate ($A$) of calcium in adult man as approximately 0.5 g per day and the total exchangeable calcium at about 5 g.

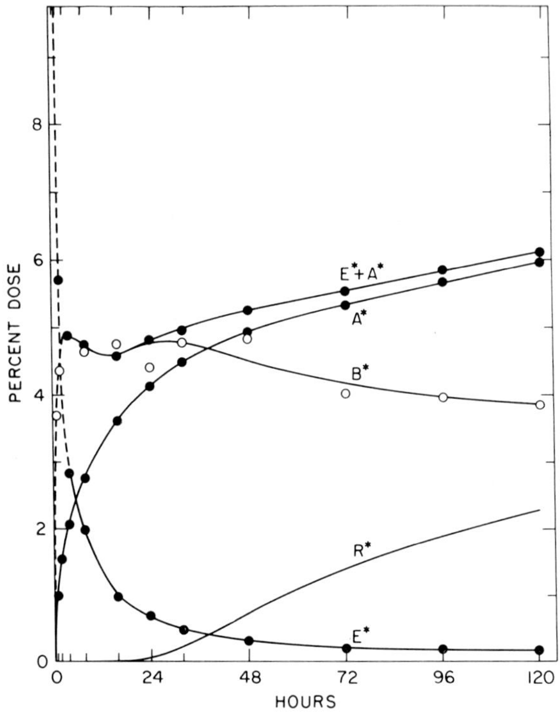


FIG. 7.2. Role of accretion, resorption, and exchange reactions in the uptake and removal of $Ca^{45}$ in the tibial ends of young rats. The data are for the same rats as in Fig. 7.1. (From Bauer *et al.*, *Mineral Metabolism*.[8])

Alternative methods[8] are described for calculation of values for $A$ and $E$ [Eq. (7.3)] and each gives a somewhat different result. A simplified method, applicable under certain circumstances and with appropriate reservations, is by extrapolation of one of the exponentials characterizing the blood disappearance curve to $t = 0$. The use made of this method by other authors is described below.

Bauer and Ray[11] studied the kinetics of $Sr^{85}$ metabolism in man. The use of strontium in tracer amounts gives data on the metabolism of bone comparable to those obtained by the use of the isotopes of calcium. Bauer and Ray

used an electronic analog computer to simulate the plasma and excretory activity curves observed during 120 hours following injection of $Sr^{85}$ in five adult men. The results of these workers are illustrated in Fig. 7.3 which represents the model used by them. The magnitudes of compartments X and Y derived from this analysis were similar to published values for the intracellular and the extravascular-extracellular spaces, respectively. Compartment Z was assumed to represent exchangeable strontium in the bone mineral; this was

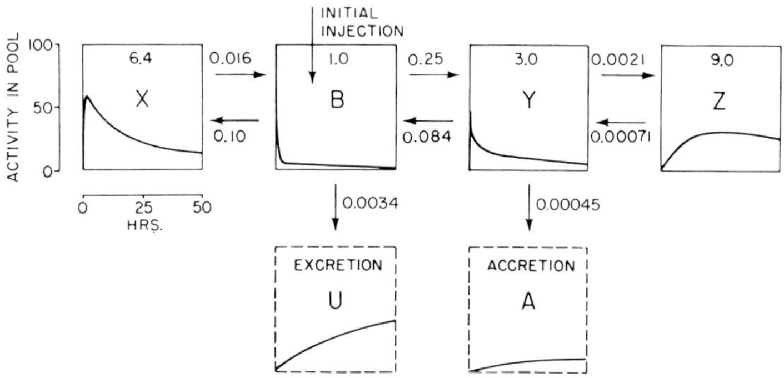


FIG. 7.3. Postulated model for strontium metabolism. Compartment B represents the vascular space (serum). The nature of compartments X, Y, and Z is unknown. The boxes U and A, containing activity drained from the system by excretion and accretion, have been outlined with dots to symbolize that they do not represent true compartments. Observed serum and excretory activity values have been simulated with an electronic analog computer. The figures within the boxes indicate relative volumes derived from the $k$ values, the volume of B being the unit of reference. For comparison, the activity-time curves for the first 50 hours following administration of activity into B have been indicated schematically. (From Bauer and Ray, *J. Bone Joint Surg.*[11])

confirmed by analysis of data obtained by external counting over bone and soft tissue.

Bauer[12] and his collaborators[13] have also pioneered in the utilization of a gamma-emitting isotope of calcium, $Ca^{47}$, in the study of the kinetics of calcium metabolism in man by employing external scanning of the skeleton with collimated scintillation counters. Bauer[14] states that each of the three isotopes $Ca^{45}$, $Ca^{47}$, and $Sr^{85}$ has a unique combination of chemical and physical properties which clearly define their applications in clinical and research work. Bauer predicts that these isotopes, as well as other bone tracers such as tetracycline, will be used in combination with one another to an increasing extent. By utilizing a two-compartment model, Wendeberg[15] has studied the kinetics of $Ca^{47}$ and $Sr^{85}$ in man, calculating accretion rate and exchangeable calcium spaces. The method is based on calculation of the total body retention at two suitable intervals of time after injection of $Sr^{85}$ or $Ca^{47}$, when mixing

of the isotope is complete within the exchangeable mineral spaces. Wendeberg reports that with one exception all of the data obtained by external counting have proved to be amenable for simulation by summing fractions of the activities in the two compartments, S and E, and the amount of activity accreted. The one exception was in the case of a fracture in which continued rise of the accretion rate in the fracture region during the period of investigation, while the fracture callus was not in a steady state, was believed to have introduced uncertainties into the calculations. Dymling, also a collaborator of Bauer, has reported[16] on accretion and excretory clearance rates and exchangeable spaces measured with $Ca^{47}$ and $Sr^{85}$ under normal and pathological conditions; references will be made in Chapter 14 to the results reported by him.

Heaney and Whedon[17] arrived at similar results. By studying adult human subjects, they made the simplifying assumption that the miscible calcium pool, despite its theoretical complexity, behaves as a single compartment. These investigators made use of extrapolation of one of the exponentials characterizing the blood-disappearance curve, as referred to above, but bypassed a possible source of error by extrapolating only from the portion of the curve representing the post-equilibration activity data. They expressed their results in terms of bone-formation rate ($BFR$), corresponding to the accretion rate ($A$), and as the miscible calcium pool or ($E$) of Bauer, Carlsson, and Lindquist. They also introduced a term $\theta$, defined as the time required for recycling of isotope through the shortest-lived species of bone, as determined by an abrupt change in the slope of the isotope disappearance curve.

Heaney[18] bases his evaluation of calcium-tracer studies on two fundamental observations: (1) the miscible calcium pool is very much smaller than the total body calcium; and (2) the turnover of this pool is considerably less than its measurable excretory or absorptive components. These observations then led to an equally fundamental assumption, i.e., that the difference between total turnover and its detectable external components represents bone formation and resorption or an approximation thereof.

Nordin and his associates[19] have studied seven normal subjects with $Ca^{47}$, using essentially the techniques of Heaney and Whedon but differing with them and with Bauer in some of the interpretations of the data. In a subsequent communication,[20] still further refinements are introduced into the calculations.

The simplified method of extrapolation of an exponential to $t = 0$ is also used by Milhaud and Aubert.[21] When applied to radiocalcium data in patients with a moderate calcium excretion, it gives almost the same values for $A$ and $E$ as those given by a more complete analysis of the data. In a later communication, Milhaud[22] has proposed separation of the four exponentials in

the blood-disappearance curve for $Ca^{45}$ for the identification and calculation of the parameters represented by the models used by him in the study of calcium kinetics.

Heaney[23] has published a critical review on the evaluation and interpretation of calcium kinetic data in man, with special attention to the measurement of bone formation and resorption, mineral exchange between body fluids and the skeleton, turnover and exchange of various nonosseous calcium compartments, and the measurement and significance of changes in these activities in health and disease. The reading of this review is essential as a study of the subject in depth, and it is recommended to the advanced student.

## 3. Whole-Body Counting

The theory and method of whole-body counting have been reviewed in Chapter 4. This technique has the great advantage that, with multichannel pulse-height analysis, it not only gives data for total gamma-ray emanation from the body but also identifies the source of the radiation by means of resolution of the energies characteristic of the gamma-emitting radioelements. Facilities for whole-body counting have come into use (1) as research tools and (2) for the monitoring of populations for contamination with fission products, or for body burdens—such as from radium—acquired either accidentally (as by radium dial painters) or as the result of therapeutic administration.

Of greatest interest for the purposes of this chapter is whole-body counting following administration of a bone-seeking isotope for medical research or diagnosis. This technique requires a gamma emitter, with special affinity for bone, and the choice is generally between $Ca^{47}$ and $Sr^{85}$. While $Ca^{47}$ is, in theory at least, the best tracer for skeletal calcium metabolism, it has, for certain purposes, the disadvantages of multiple gamma energy peaks and of contamination with undesirable $Ca^{45}$.

MacDonald[24] has explored the use of $Sr^{85}$ as a bone seeker for whole-body counting. This isotope of strontium approximates the behavior of calcium in so far as skeletal metabolism is concerned. It has the advantages of gamma rays solely of 0.51 Mev and of a half-life of 65 days which permits its detection by whole-body counting for a long period after a single dose. Thus, by using the total body counter, the retention of $Sr^{85}$ in a patient with osteoporosis was found to be 0.3% of the dose *one year* after administration of the isotope.

In assessing the kinetics of the bone mineral with $Sr^{85}$ or $Ca^{47}$, it is often necessary to secure data for the amount remaining in the body at intervals following administration. For this purpose, the usual procedure is to subtract the amounts excreted in the urine and the feces from the dose, but ob-

taining data in this fashion may be laborious and time consuming. In comparison, similar data may be obtained by whole-body counting in a matter of minutes. Table 7.1 shows the cumulative excretion of $Sr^{85}$ over a period of 21 days following a single intravenous injection, as observed by counting of urine and feces and as calculated from total body count.

TABLE 7.1
CUMULATIVE EXCRETION OF STRONTIUM-85[a,b]

| Day | Total per cent of dose in excreta: Urine | Feces | Urine + feces | Calculated from total body count |
|---|---|---|---|---|
| 4 | 44.7 | 1.0 | 45.7 | 45 |
| 7 | 52.7 | 9.2 | 61.9 | 61 |
| 13 | 59.2 | 12.8 | 72.0 | 71 |
| 18 | 61.6 | 13.6 | 75.2 | 73 |
| 21 | 62.4 | 13.9 | 76.3 | 73 |

[a] From MacDonald, *USAEC Report*.[24]
[b] Following a single intravenous injection.

The clinical applications of whole-body counting of bone-seeking isotopes will be reviewed in Chapter 14. The above discussion will serve to illustrate the potentialities of the method since the procedure has not yet come into use sufficiently to permit a complete evaluation of its usefulness.

## 4. Compartmental Analysis

A complete review of the theory and use of tracers in compartmental analysis will not be given here. Such reviews have already been published by Robertson[25] and by Solomon.[26] Compartmental analysis of biological systems, as described by Solomon, is concerned primarily with the development of abstract models which reproduce the kinetics of tracer distribution. Such models are, of course, only incomplete representations of whole biological systems, and the abstractions in the models should thus only be partially identified with morphological or functional components of the system.

An early use of compartmental analysis in the study of calcium metabolism was that of Maletskos.[27] In this work, the compartments were treated solely as abstractions, and no attempt at identification was made. Cohen *et al.*[28] made some use of the principles of compartmental analysis in tracer studies of bone grafts. A further advance was made by Bauer and Ray[11] who employed an electronic analog computer for simulation of the transfer of labeled ions from compartment to compartment in a four-compartment model and for the calculation of rate constants. One compartment was identified as the vascular

space (serum); the remaining compartments were not positively identified but were assumed to represent intracellular and extracellular spaces and the exchangeable strontium in the bone mineral. Compartmental analysis is now recognized as a useful tool in the study of skeletal metabolism and will doubtless find additional applications.

## 5. The Power Function

As indicated above, the kinetics of the bone mineral has generally been studied by the use of exponential functions in which time $t$ appears in the exponential. A similar use of the exponential relation is described in Chapter 1 to illustrate the kinetics of the decay of radioactive elements. The exponential function was characterized by Lord Kelvin as the law of compound interest, by means of which interest is compounded continuously instead of at arbitrary intervals. An exponential function appears as a straight line when plotted on semilog graph paper.

A second method used in the study of the metabolism of the alkaline earth radioisotopes is the *power function.* This is an expression in which the independent variable is raised to a constant power. For present purposes, this has the form $t^{-b}$, where $t$ is the time and $b$ is a positive fraction between zero and one. When such an expression is plotted *versus* time on log-log graph paper, the result is a straight line.

Norris *et al.*[29] applied the power function to the retention of radioactive bone seekers and found that it provides a satisfactory empirical description of the retention of several radioactive isotopes preferentially deposited in skeletal tissue. Tyler[30] had previously introduced a modification of the power function to prevent its failure at very short times after administration of the isotope when it increases without limit as $t$ approaches zero.

Marshall[31] has extended the usefulness of the power function by showing its general applicability to the study of the metabolism of the bone-seeking alkaline earth radioisotopes with the use of only three adjustable parameters—accretion rate, excretion rate, and the very small constant added to $t$ as a modification of the function in its original form. Figure 7.4 illustrates a log-log graph of the specific activities of the whole body and of the serum in the dog for a period of 3 years following a single injection of $Ra^{226}$. The most striking aspect of the power function model, as well as of the data, is that the serum-activity curve crosses the whole body curve at about 45 days but that the straight-line relation for both is maintained for the 3 years during which observation was continued. This *crossover time,* $t_x$, appears to be a useful concept and should be helpful in understanding the experimental data.

From observations on individuals carrying body burdens of radium, it

has been found that retention follows a power function for many years, during which the specific activity curves continue as straight lines. Eventually, however, a radioisotope reaches a steady state within the body-blood system, and subsequent to this time retention follows an exponential function; the time at which this occurs may be calculated from data previously obtained.

Marshall has also applied the power-function analysis to the Bauer, Carlsson, and Lindquist formulation,[9] rephrasing the basic assumptions to read that the activity of a radioisotope in the body can, at any time following its intro-

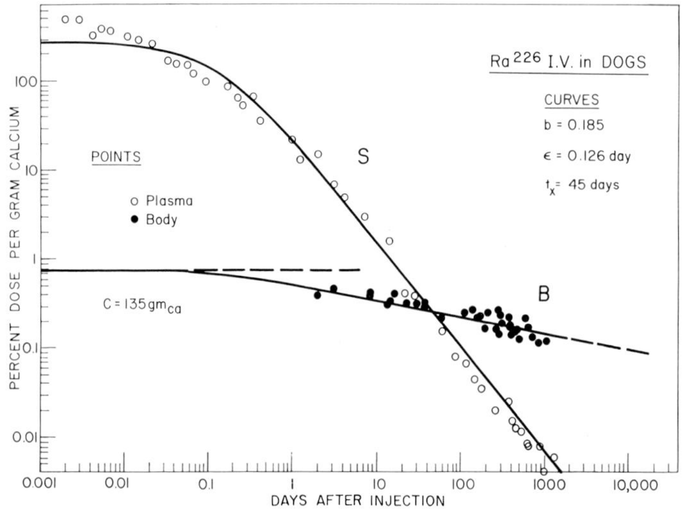


FIG. 7.4. Log-log graph of specific activities of whole body and of serum of dogs, for 3 years following a single injection of $Ra^{226}$. Data from several groups of beagles about 1.3 years of age at the time of the intravenous injection. (From Marshall, *USAEC Report*.[31] Recalculated from Van Dilla *et al.*, *Radiation Res.*[32])

duction into the serum, be separated conceptually into two fractions, the first changing in proportion to the serum specific activity, $S$, and the second changing in proportion to the time integral of $S$. In mathematical form, this may be written as

$$qR = ES + A \int_0^t S(t)\, dt \tag{7.5}$$

in which $q$ is a quantity of radioactivity in microcuries introduced into the serum at time $t = 0$, $R$ is the fractional retention of this activity in the body, $E$ is a quantity of calcium in grams, and $A$ is a rate of transfer (grams of calcium per day). All capital letters refer to functions of time ($t$). This equation is identical with Eq. (7.3) but employs different symbols. Since $A$ has usually been measured between 3 and 7 days after administration of the

isotope, it may be designated as $A_5$, which is a useful parameter for the characterization of the metabolic state of a skeletal system. According to these interpretations, $E_5$ is a measure of the calcium content of all those deposits with terms of less than 5 days, while $A_5$ is a measure of the rate of uptake from the serum of calcium destined for deposits with terms of more than 5 days.

While power function has usually been described as an empirical formulation, in contrast to the exponential function which represents the compound interest law, Marshall has formulated a postulate (to be regarded as speculative for the present) which states that the power function is a statistical property of any sufficiently complex system which has evolved to a state of dynamic equilibrium. Such a postulate, if true, would provide a rational rather than an empirical basis for interpretation of systems studied by means of the power function and might give added meaning to this method for the study of the kinetics of skeletal metabolism.

## References

1. O. Chiewitz and G. Hevesy, Radioactive Indicators in the Study of Phosphorus Metabolism in Rats, *Nature (London)*, **136**: 754-755 (1935).
2. W. P. Norris and W. Kisieleski, Comparative Metabolism of Radium, Strontium, and Calcium, *Cold Spring Harbor Symp. Quant. Biol.*, **13**: 164-172 (1948).
3. W. D. Armstrong, J. A. Johnson, L. Singer, R. I. Lienke, and M. L. Premer, Rates of Transcapillary Movement of Calcium and Sodium and of Calcium Exchange by the Skeleton, *Am. J. Physiol.*, **171**: 641-651 (1952).
4. R. O. Thomas, T. A. Litovitz, M. I. Rubin, and C. F. Geschickter, Dynamics of Calcium Metabolism. Time Distribution of Intravenously Administered Radiocalcium, *Am. J. Physiol.*, **169**: 568-575 (1952).
5. R. O. Thomas, T. A. Litovitz, and C. F. Geschickter, Alterations in Dynamics of Calcium Metabolism by Intraintestinal Calcium Reservoirs, *Am. J. Physiol.*, **176**: 381-387 (1954).
6. S. L. Hansard, C. L. Comar, and G. K. Davis, Effects of Age Upon the Physiological Behavior of Calcium in Cattle, *Am. J. Physiol.*, **177**: 383-389 (1954).
7. F. Bronner, R. S. Harris, C. J. Maletskos, and C. E. Benda, Studies in Calcium Metabolism. The Fate of Intravenously Injected Radiocalcium in Human Beings, *J. Clin. Invest.*, **35**: 78-88 (1956).
8. G. C. H. Bauer, A. Carlsson, and B. Lindquist, Metabolism and Homeostatic Function of Bone, in *Mineral Metabolism*, Vol. I, pp. 609-676, Academic Press Inc., New York, 1961.
9. G. C. H. Bauer, A. Carlsson, and B. Lindquist, Evaluation of Accretion, Resorption, and Exchange Reactions in the Skeleton, *Kungl. Fysiol. Sällskapets Lund Förhandl.*, **25**: 1-16 (1955).
10. G. C. H. Bauer, A. Carlsson, and B. Lindquist, Use of Isotopes in Clinical Studies of Skeletal Metabolism, in *Radioaktive Isotope in Klinik und Forschung*, III, pp. 25-39, Urban & Schwarzenberg, München, 1958.
11. G. C. H. Bauer and R. D. Ray, Kinetics of Strontium Metabolism in Man, *J. Bone Joint Surg.*, **40-A**: 171-186 (1958).

12. G. C. H. Bauer, Experience with External Counting Techniques in Clinical Studies of Bone, in *Medical Uses of* $Ca^{47}$, IAEA Technical Reports Series No. 10, pp. 92-95, Intern. Atomic Energy Agency, Vienna, 1962.
13. B. Wendeberg, Mineral Metabolism of Fractures of the Tibia in Man Studied with External Counting of $Sr^{85}$, *Acta Orthopaed. Scand., Suppl.*, **52** (1961).
14. *Medical Uses of* $Ca^{47}$, IAEA Technical Reports Series No. 10, Intern. Atomic Energy Agency, Vienna, 1962.
15. B. Wendeberg, Kinetics of $Ca^{47}$ and $Sr^{85}$ in Man, in *Medical Uses of* $Ca^{47}$, IAEA Technical Reports Series No. 10, pp. 51-57, Intern. Atomic Energy Agency, Vienna, 1962.
16. J. F. Dymling, Accretion and Excretory Clearance Rates and Exchangeable Spaces Measured in Man with $Ca^{47}$ and $Sr^{85}$ Under Normal and Pathological Conditions, in *Medical Uses of* $Ca^{47}$, IAEA Technical Reports Series No. 10, pp. 73-76, Intern. Atomic Energy Agency, Vienna, 1962.
17. R. P. Heaney and G. D. Whedon, Radiocalcium Studies of Bone Formation Rate in Human Metabolic Bone Disease, *J. Clin. Endocrinol. Metab.*, **18**: 1246-1267 (1958).
18. R. P. Heaney, Evaluation of Calcium Kinetics in Adult Humans, in *Medical Uses of* $Ca^{47}$, IAEA Technical Reports Series No. 10, pp. 48-51, Intern. Atomic Energy Agency, Vienna, 1962.
19. B. E. C. Nordin, M. Bluhm, and J. MacGregor, *In vitro* and *in vivo* Studies with Bone-seeking Isotopes, in *Radioisotopes and Bone*, pp. 105-126, Blackwell Scientific Publ., Oxford, 1962.
20. B. E. C. Nordin and J. MacGregor, Analysis of Methods for Interpretation of Tracer Data on Bone, in *Medical Uses of* $Ca^{47}$, IAEA Technical Reports Series No. 10, pp. 57-61, Intern. Atomic Energy Agency, Vienna, 1962.
21. G. Milhaud and J. P. Aubert, Étude des Voies Générales du Métabolisme Calcique chez l'Homme, in *Radioisotopes in Scientific Research*, Vol. IV, pp. 120-124, Pergamon Press, New York, 1958.
22. G. Milhaud and J. P. Aubert, Studies on the Intestinal Absorption of Dietary Calcium in Man with $Ca^{45}$, in *Medical Uses of* $Ca^{47}$, IAEA Technical Reports Series No. 10, pp. 39-42, Intern. Atomic Energy Agency, Vienna, 1962.
23. R. P. Heaney, Evaluation and Interpretation of Calcium-kinetic Data in Man, *Clin. Orthopaed.*, **30** (1963).
24. N. S. MacDonald, *Recent Uses of a Total Body Counter Facility in Metabolic Research and Clinical Diagnosis with Radionuclides*, USAEC Report UCLA-478, pp. 1-28, University of California at Los Angeles, June 1961.
25. J. S. Robertson, Theory and Use of Tracers in Determining Transfer Rates in Biological Systems, *Physiol. Rev.*, **37**: 133-154 (1957).
26. A. K. Solomon, Compartmental Methods of Kinetic Analysis, in *Mineral Metabolism*, Vol. I, pp. 119-168, Academic Press Inc., New York, 1960.
27. C. J. Maletskos, *The Dynamics of Calcium Metabolism*, Thesis, Massachusetts Institute of Technology, Cambridge, 1954.
28. J. Cohen, C. J. Maletskos, J. H. Marshall, and J. B. Williams, Radioactive Calcium Tracer Studies in Bone Grafts, *J. Bone Joint Surg.*, **39-A**: 561-577 (1957).
29. W. P. Norris, S. A. Tyler, and A. M. Brues, Retention of Radioactive Bone-seekers, *Science*, **128**: 456-462 (1958).
30. S. A. Tyler, *On a Modification of the Power Function. Description of Body Burden from Measured Retention*, USAEC Report ANL-5841, pp. 132-138, Argonne National Laboratory, March, 1958.

31. J. H. Marshall, *Theory of Alkaline Earth Metabolism*, USAEC Report ANL-6646, pp. 5-29, Argonne National Laboratory, February, 1963.
32. M. A. Van Dilla, B. J. Stover, R. L. Floyd, D. R. Atherton, and D. H. Taysum, Radium ($Ra^{226}$) and Radon ($Em^{222}$) Metabolism in Dogs, *Radiation Res.*, **8**: 417-438 (1958).

# Chapter 8

# *Mineral Metabolism*

## 1. The Internal Environment of Bone

Bone tissue shares an internal environment with the other tissues of the body. This environment includes the circulating blood plasma and the intercellular fluid—a total of approximately 20% of the weight of the body. An even larger amount of fluid, 50% of the body weight, is intracellular; a small proportion of this is in the cellular elements of bone.

There is an extremely rapid exchange of water and dissolved substances among the various compartments of the body which serves not only to bring oxygen to and take carbon dioxide away from the cells but also to transfer and exchange fluids and electrolytes. It is by the movement of water and dissolved substances that the minerals essential to the bones are brought to the locations at which they are to be deposited and are released when needed to maintain the physiological constancy of the composition of the body fluids.

The important cation of the internal environment, from the standpoint of bone, is calcium. Of this element, a 70-kg man has in the circulating plasma approximately 280 mg of which half, or 140 mg, is in the form of calcium ions. This calcium does not exchange to any considerable degree with the contents of cells, which are virtually free from calcium, but it does exchange freely and rapidly with the 10 liters of intercellular fluid which contains about 500 mg of calcium, nearly all in the ionized form. In adult man one calcium ion out of four leaves the blood plasma every minute.[1] The plasma plus the intercellular fluid forms the calcium compartment which amounts to 15–20% of the weight of the body and contains, in a 70-kg man, 525–700 mg of calcium ions.

This amount of calcium, in the ionized form, is seemingly insignificant when compared with 1200 g of calcium in the skeleton. Moreover, in view of the very rapid exchange of calcium between the fluid-calcium compartment and the bones, it is remarkable that the organism is able, by means of homeostatic mechanisms, to maintain a constant level of calcium ions in the body fluids, especially when an individual ion may stay in the plasma for only a matter of minutes.

Of the anions in the fluids of the body, the most important for the maintenance of osmotic pressure are $Cl^-$ and $HCO_3^-$. From the standpoint of the mineral constituents of bone, the most important is inorganic phosphate which

is mainly present as the ions $HPO_4^{2-}$ and $H_2PO_4^-$. Since these ions exchange to a considerable degree with the contents of cells, the space available to phosphate in the body is larger than the calcium space, but the same considerations with respect to the rapidity of exchange are valid also for phosphate ions.

## 2. Calcium of the Fluids of the Body

All or nearly all of the calcium in the blood is in solution in the plasma; the amount in the red blood cells is negligible. In normal man the concentra-

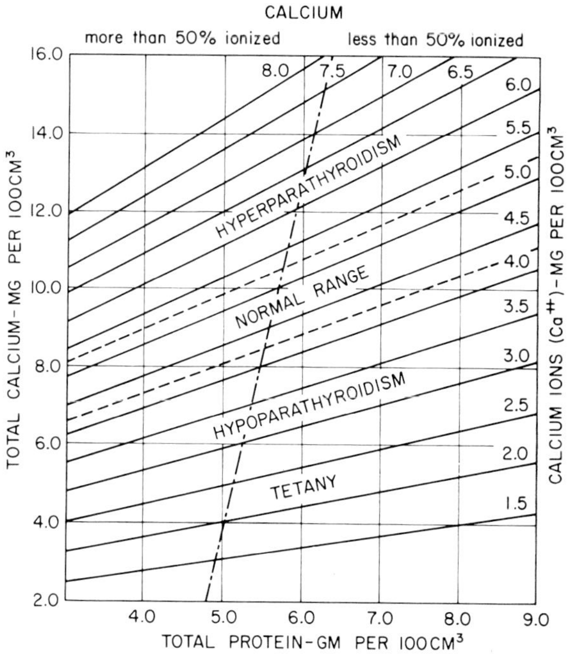


FIG. 8.1. Nomogram for calculation of $Ca^{2+}$ concentration from total protein and total calcium of serum or plasma. Constructed from mass-law equation $[Ca^{2+}] \times [Prot^{2-}]/[CaProt] = K$. (From McLean and Hastings, *Am. J. Med. Sci.*[2])

tion of calcium in the plasma is usually between 9 and 11 mg/100 cm³, with 10 mg/100 cm³ (2.5 mM per liter) a representative figure. This is, as indicated above, in constant exchange with the calcium of the extracellular fluid and that of the bones; the regulation of the level of calcium in the circulating fluid will be considered in Chapter 9.

It has been known since early in the present century that the calcium in the plasma is separable into two major and roughly equal fractions—diffusible

and nondiffusible. It has also been known that the nondiffusible fraction is associated with the plasma proteins. The relationship of calcium to the plasma proteins has been incorporated in a formulation which describes the plasma as

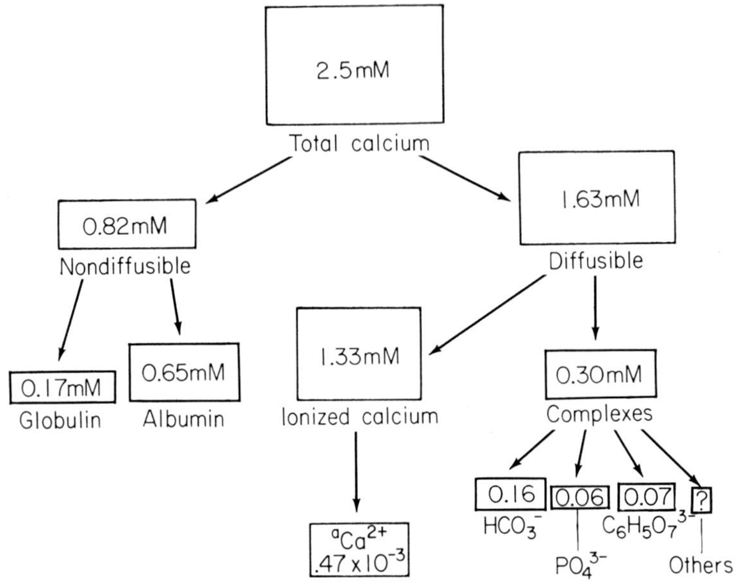


FIG. 8.2. The state of calcium in normal serum as calculated from ultrafiltration data and formation constants. mM designates mM/liter. (From Neuman and Neuman, *The Chemical Dynamics of Bone Mineral*.[3])

a solution of a weak electrolyte, calcium proteinate, the ionization of which may be represented by the equation

$$\mathrm{CaProt} \leftrightarrows \mathrm{Ca}^{2+} + \mathrm{Prot}^{2-} \tag{8.1}$$

and by the mass-law equation

$$\frac{[\mathrm{Ca}^{2+}] \times [\mathrm{Prot}^{2-}]}{[\mathrm{CaProt}]} = K_{\mathrm{CaProt}} = 10^{-2.22} \tag{8.2}$$

From these two equations, the $Ca^{2+}$ concentration of the plasma may be calculated from analyses for total calcium and total protein—quantities easily determined in the laboratory. The calculation is facilitated by the use of a nomogram,[2] shown in Fig. 8.1, which illustrates the relationships of the ionized and un-ionized fractions of calcium at varying protein and calcium levels. At normal protein levels, slightly more than half of the plasma calcium is in the ionized form; the remainder is mainly in association with protein. The distribution[3] of the total calcium in the plasma is illustrated in Fig. 8.2.

In the protein-poor fluids of the body, such as the cerebrospinal and the

intercellular fluids, the state of calcium approximates that in an ultrafiltrate of plasma. In such fluids, and this may be taken to apply to the extracellular fluids of bone, the concentration of calcium is generally about 5 mg/100 $cm^3$ (1.25 mM per liter), virtually all in the ionized form.

In calcification, when calcium is to be deposited in the bones, it moves first from the plasma to the intercellular fluid and then is deposited in locations prepared for calcification. Under such circumstances, the plasma requires replenishment in order to maintain $Ca^{2+}$ concentrations at physiological levels. Such replenishment of calcium ordinarily comes from the diet, but the skeleton itself is the principal depot for the storage of reserve supplies in the organism. Thus it frequently comes about that, as calcium is being deposited in one or more places in the skeleton, it is being withdrawn from other places; the entire system *plasma-intercellular fluid-bone* remains in a steady state.

## 3. Phosphates of the Fluids of the Body

Inorganic phosphate is present in the fluids of the body in the form of the ions of orthophosphoric acid, $H_3PO_4$. The transport of phosphate within the animal organism is a function of the circulating blood and of the intercellular fluid. Since the phosphates have a manifold role in the organism and are closely related to many metabolic functions, as well as to calcification, they are found in many forms and in many locations. It is, in fact, this feature of the metabolism of phosphorus that limits the usefulness of radiophosphorus, $P^{32}$, as a tracer element. The movement of phosphate within and between the fluid compartments of the body and between fluid and bone occurs constantly and with the same rapidity as the movement of calcium.

Of approximately 700 g of phosphorus in the adult human body, about 600 g are in the skeleton. The presence of this large amount of phosphorus in the bones depends upon the ability of orthophosphoric acid to form difficultly soluble compounds with calcium. Orthophosphoric acid is a weak tribasic acid. Its dissociation constants, corrected for the ionic strength of serum,[4] are given in Table 8.1. From these constants it may be calculated that at pH 7.4, approximately 85% of the total inorganic phosphate of the plasma

TABLE 8.1

DISSOCIATION OF $H_3PO_4$[a]

| Reaction | Constant | pK |
|---|---|---|
| $H_3PO_4 \rightleftharpoons H^+ + H_2PO_4^-$ | $K'_1 = 1.22 \times 10^{-2}$ | $pK'_1 = 1.915$ |
| $H_2PO_4^- \rightleftharpoons H^+ + HPO_4^{2-}$ | $K'_2 = 2.19 \times 10^{-7}$ | $pK'_2 = 6.66$ |
| $HPO_4^{2-} \rightleftharpoons H^+ + PO_4^{3-}$ | $K'_3 = 1.66 \times 10^{-12}$ | $pK'_3 = 11.78$ |

[a] Derived from data of Sendroy and Hastings, *J. Biol. Chem.*[4]

is in the form of the divalent ion $HPO_4^{2-}$ while 15% is monovalent $H_2PO_4^-$ and only 0.0035% is trivalent $PO_4^{3-}$.

## 4. Physiological Factors in Distribution of Ions

Physiological factors in mineral metabolism, with more or less relationship to the physiology of bone, have been studied by tracer methods. Of these factors the most important are: (1) transfer of ions across biological membranes; (2) tubular reabsorption in the kidneys; (3) absorption in the gastrointestinal tract; and (4) transfer by the mammary glands in the secretion of milk. The influence of vitamins and hormones on all of these factors has been a subject of intensive study. A further subject of inquiry has been the movement of mineral ions, especially calcium, into muscle and nerve tissue.

## 5. Discrimination Factors

Of especial current interest is the discrimination against strontium, as compared with calcium, in its passage through the animal organism. This discrimination occurs initially in the animals that consume radioactive strontium, in which the intestinal absorption of strontium is relatively less than that of calcium. Secretion of strontium in the milk is less than that of calcium, and reabsorption of strontium in the renal tubules is also less. As a result the animals' consumption of plant food, which may either have strontium incorporated within its structure or may have it on its surfaces, reduces the hazard to man of fallout even before man is exposed to it. After man consumes strontium, the same factors operate to provide further protection—i.e., discrimination in intestinal absorption and in tubular reabsorption. Further protection of the fetus and infant is provided by discrimination against strontium in passage across the placenta and in secretion of milk.

Comar and Wasserman[5] have proposed the term Observed Ratio (*OR*) to represent the over-all behavior of Sr and Ca between some given sample and a precursor, and they have reserved the term Discrimination Factor (*DF*) to apply to a particular physiological process that brings about the differential behavior. The following formulas are representative of the use of these terms:

$$OR_{\text{sample-precursor}} = \frac{\text{Sr/Ca of sample}}{\text{Sr/Ca of precursor}}$$

Discrimination factors—

$$OR = (DF_1)\ (DF_2)\ (DF_3) \ldots (DF_n)$$

From the standpoint of this volume, the Observed Ratios of most interest are those from diet to bone. $OR_{\text{diet to bone}}$ has been determined on various diets

in a variety of animals and in man and by employing a variety of methods in combination with both stable and radioactive elements. The *OR* values, Sr/Ca, tend to fluctuate between 0.25 and 0.50 in man, as well as in other animals, and the discrimination between strontium and calcium between diet and bone depends mainly upon the $DF_{\text{diet-blood}}$ and $DF_{\text{blood-urine}}$. There is little evidence, from single-tracer experiments, of differences in rate of movement of Ca and Sr between blood and bone, and discrimination at this point, if it exists at all, appears not to be physiologically important. As noted elsewhere, $Sr^{85}$ in tracer amounts gives information concerning skeletal uptake comparable to that obtained by the use of $Ca^{45}$ or $Ca^{47}$.

The application of these observations to the hazard to man from $Sr^{90}$ in radioactive fallout may be summarized as follows. When $Sr^{90}$ is released from the biosphere, it is deposited both on the surfaces of plants and on the soil; relatively little may be expected to reach the organism directly by inhalation. Plants take it up from the soil, and there is, in addition, foliar absorption of the element and perhaps some unabsorbed residue on the leaves and stems of the plants. For the most part, the hazard is in the food chain, where it moves from plants to animals. Contaminated plants may also be consumed directly by man. The amounts of $Sr^{90}$ that may be expected to reach man in drinking water or through edible fish may be safely ignored for present purposes.

In our civilization, of the farm animals, the dairy cow supplies by far the greater proportion of $Sr^{90}$ that may reach man in the food chain. If the discrimination factor, soil to milk, is added to the further discrimination factors in man, following ingestion of milk, it is seen that the hazard to man of $Sr^{90}$ in fallout is reduced successively at each stage. The most important index of the hazard to man is the *rate of deposition* of $Sr^{90}$ from the biosphere to the earth; the earlier assumption that the cumulative deposit was the primary factor determining the contamination of milk, and hence the hazard to man, has now been largely abandoned. Wasserman *et al.*[6] have reported on placental discrimination against strontium in the rat and have published data for Observed Ratios and Discrimination Factors, included in Table 8.2. Placental discrimination, together with discrimination in the secretion of

TABLE 8.2

PLACENTAL DISCRIMINATION IN THE RAT[a]

| | | | |
|---|---|---|---|
| $OR_{\text{mother-diet}}$ | $= 0.28$ | $OR_{\text{mother-urine}}$ | $= 4.7$ |
| $OR_{\text{fetus-diet}}$ | $= 0.17$ | $OR_{\text{fetus-mother}}$ | $= 0.71$ |
| $DF_{\text{absorptive}}$ | $= 0.42$ | | |
| $DF_{\text{urinary}}$ | $= 0.63$ | | |
| $DF_{\text{placental-fetus}}$ | $= 0.65$ | | |
| $DF_{\text{placental-dam}}$ | $= 1.22$ | | |

[a] From Wasserman *et al.*, *Am. J. Physiol.*[6]

milk, combine to afford a considerable degree of protection to the fetus and infant.

## 6. Transfer of Ions across Biological Membranes

Ions, including both cations and anions, move in one direction or another across biological membranes of different sorts and under different circumstances. In its simplest form, ion transfer across a membrane, such as a cell membrane, occurs as diffusion which requires no transfer of energy and which is usually in both directions. In another form of transfer which involves movement across a cell membrane, concentrations of ions inside and outside the cell are maintained against a concentration gradient. An outstanding example is the difference in concentrations of sodium and of potassium inside the cell, e.g., a muscle cell or red blood cell, as compared with their concentrations in the fluids bathing the cell. In such instances, there is postulated some sort of an ion pump that provides the energy necessary to oppose the gradient which tends toward equilibrium; such a pump is usually assumed to obtain energy from intracellular glycolysis. In still another form of transfer, in which the membrane is not a single layer of cells forming a passive barrier, there is now much evidence for active transport, which is defined as transport across a membrane, against a concentration gradient and an electropotential gradient, both requiring energy transfer.

The transfer of calcium and of strontium across the intestinal wall, across the renal tubular epithelium, across the placental barrier, and across the alveolar epithelium of the lactating mammary glands is regarded as active transport. With the exception of the intestinal wall, which both absorbs and secretes calcium, such transfer is unidirectional under physiological conditions.

The active transport of $Ca^{45}$ by the small intestine has been studied in detail by Schachter and associates[7-9] by the use of everted gut-sacs prepared from segments of the small intestine of young rabbits, rats, and guinea pigs. Such gut-sacs, used in *in vitro* studies, transport $Ca^{45}$ from the mucosal to the serosal surfaces against concentration gradients and thus fulfill the requirements for active transfer. Early reports indicated that (1) active transport mechanism is limited in capacity, (2) is dependent on oxidative phosphorylation, and (3) appears to be relatively specific for $Ca^{2+}$ and $Mg^{2+}$ (in contrast to $Sr^{2+}$ and $Ba^{2+}$). The mechanism is also dependent on pretreatment with vitamin D of the animals from which the gut-sacs are prepared, and it was reported that neither the transport of $Ca^{45}$ nor the effects of vitamin D thereon could be explained by assuming that citrate acted as a carrier. Active transport was inhibited by such metabolic poisons as sodium cyanide and 2,4-dinitrophenol.

These *in vitro* observations have been supported by further reports by

Rasmussen,[10] in studies on the effect of the parathyroids on calcium transport; by Wasserman,[11] in studies of the metabolic basis of calcium and strontium discrimination; and by Harrison *et al.*,[12] in studies of the effect of vitamin D on intestinal absorption of calcium. Wasserman *et al.*[13] have found, in the living rat, that ionic calcium is transferred against a concentration gradient and an electropotential gradient by the duodenal membrane, and they regard this as further confirmation of the *in vitro* studies.

The everted gut-sac *in vitro* is adapted to the observation of unidirectional movement of calcium from the mucosal to the serosal surface of the preparation; it is not believed to move calcium from the serosal to the mucosal surface. The gut-sac does not, therefore, provide a completely satisfactory model of the conditions in an intact animal, in which calcium moves both from the lumen of the intestine to the blood plasma and from the plasma to the intestinal lumen. The *in vivo* preparations of Wasserman *et al.*,[13] however, permit estimates of efflux, influx, and net efflux (or net transfer) and approximate the conditions in the intact animal.

The conclusion that the active transport mechanism is not effective for $Sr^{2+}$ and $Ba^{2+}$ is based on the observation that transport of these ions from the mucosal to the serosal surfaces of everted gut-sacs was not inhibited by the agents which depress the transfer of $Ca^{2+}$ and of $Mg^{2+}$. In the experiments of Wasserman in which surviving intestinal segments were used *in vitro*, $Ca^{2+}$ but not $Sr^{2+}$ was transported against a concentration gradient from mucosa to serosa in duodenal segments while $Sr^{85}$ but not $Ca^{45}$ was transferred against concentration gradients from serosa to mucosa in ileal segments.

In view of the above observations, mainly under nonphysiological conditions and in part conflicting, recourse may be had to balance experiments with intact animals and to the experiments of Lengemann and Comar.[14] When $Sr^{85}$ and $Ca^{45}$ were injected simultaneously into ligated sections of the small intestines of rats, more radiocalcium than radiostrontium was absorbed from the intestine. In the same experiments, more of the absorbed radiostrontium than of radiocalcium was deposited in bone; this effect could be neutralized by injection of lactose, lysine, or glucose into the ileum together with the ions of the alkaline earths. The significance of the differential uptake by bone is in doubt.

## 7. Absorption and Excretion of Calcium

The absorption of calcium takes place chiefly in the upper part of the small intestine, in both the duodenum and the jejunum. A large and variable part of the ingested calcium passes through the alimentary canal unabsorbed. Moreover, a considerable quantity of calcium, having been absorbed, is se-

creted with the digestive juices. The greater part of this amount is reabsorbed while a smaller portion appears in the feces. From the standpoint of *net absorption,* normal adults may utilize as little as 20% of the calcium ingested. In *rickets* the net absorption is greatly reduced and may even fall to negative values which indicate a loss rather than a gain. Negative calcium balances are also frequently observed in otherwise healthy adults and may in part be due to variations in intake.

Nicolaysen and Eeg-Larsen[15] have distinguished *exogenous calcium,* which passes through the intestinal tract and is excreted without having been absorbed, from *endogenous calcium,* which having been absorbed from the intestines into the blood stream may be secreted into the digestive juices. Such endogenous calcium may of course be reabsorbed, and absorption and resecretion may occur any number of times. The fecal calcium is then a mixture of exogenous with such endogenous calcium as may find itself lost to the organism by fecal excretion. Nicolaysen and Eeg-Larsen have also denied the older idea of excretion of calcium by the lower bowel, and current concepts are that there is no regulated excretion of calcium by way of the gastrointestinal tract.

The most important single factor in the absorption of calcium is vitamin D whose part in homeostatic regulation of the concentration of calcium in the blood plasma is given consideration in Chapter 9. In addition, Nicolaysen[16] in 1943 introduced the concept of an *endogenous factor* in the *regulation of calcium absorption* in growing animals and suggested that the immature skeleton, possibly by the production of a hormone, influences the rate at which calcium passes through the intestinal epithelium. Nicolaysen also found that this factor has little if any effect in the absence of vitamin D and stated that in adults vitamin D exerts only a slight effect on the absorption of calcium because the endogenous factor is absent. Again, in 1953, Nicolaysen *et al.*[17] emphasized the striking correlation between body saturation with calcium and the rate of its absorption and repeated the observation that the endogenous factor responsible for this correlation is active only in the presence of vitamin D. Later, Haavaldsen *et al.*[18] concluded that the absorption of calcium throughout life is subject to regulation by vitamin D and an endogenous factor. In 1953 Carlsson[19] reported on experiments in which he used radiocalcium to study the interrelationships between vitamin D and dietary calcium and phosphorus and stated that his experiments might be considered to support the view that the absorption of calcium is controlled in some way by the bones. The idea of an endogenous factor which acts to transmit information resulting in regulation of absorption of calcium in accordance with the needs of the skeleton, while attractive, still rests on indirect evidence.

Loss of calcium from the body takes place through two main channels,

the alimentary canal and the urinary tract. Under ordinary conditions, the calcium in the feces is greater in amount than that of the urine, and under conditions of low intake may exceed that in the food. The effect of increased intake of calcium upon net absorption is variable. Most healthy adults are able to adapt to low intakes of calcium, provided that there is no deficiency of vitamin D, and this adaptation is part of the evidence for an endogenous factor. Moreover, while extremely low intake of calcium (less than 200 mg daily) may lead to continuous negative balances, there is no agreement as to when this results in an unphysiological condition, as long as the normal plasma calcium level is maintained.

## 8. Differential Excretion

Bauer *et al.*[1] studied the distribution and excretion of $Sr^{90}$ and $Ca^{45}$ in young rats at varying intervals after intraperitoneal injection of the isotopes. Radiostrontium was retained to a lesser extent than radiocalcium, possibly owing to the preferential excretion or failure of tubular reabsorption of strontium in the kidneys as well as to discrimination in favor of absorption of calcium in the intestines. Bauer and associates found no evidence for a difference between the two elements in single-tracer experiments with respect to skeletal accretion and exchange and concluded that the skeleton was unable to distinguish between these two alkaline earths; this applied to the tracer amounts of thc isotopes used in their experiments. Comar *et al.*[23] also reported a preferential excretion of strontium in the urine and concluded that this might account in considerable measure for the over-all preferential utilization of calcium over strontium. By utilizing $Sr^{85}$ and $Ca^{45}$ in man, Spencer *et al.*[20] found the renal clearance of strontium to be significantly higher than that of calcium; this condition, together with the more efficient absorption of radiocalcium, led to a retention of strontium of only one-fourth to one-half of that of calcium. All evidence is to the effect that the strontium present in the glomerular filtrate is less well reabsorbed into the circulation than is calcium and that this accounts for the discrimination factor, blood to urine.

## 9. Differential Secretion

Because of the importance of dairy milk in the food chain, soil to man, the secretion of strontium in milk as compared with that of calcium has been extensively studied. Comar and Wasserman[5] have reviewed the data obtained by double-tracer studies ($Sr^{89}$ and $Ca^{45}$) in cows and goats, which reveal a consistently lower secretion of strontium in milk and a consistently higher excretion in feces and urine, as compared with calcium. Table 8.3 illustrates

TABLE 8.3

SUMMARY OF OBSERVED RATIO VALUES IN THE LACTATING COW[a]

| Treatment | Cow no. | $OR_{\text{plasma-diet}}$ | $OR_{\text{urine-diet}}$ | $OR_{\text{urine-plasma}}$ | $OR_{\text{feces-diet}}$ | $OR_{\text{milk-diet}}$ |
|---|---|---|---|---|---|---|
| Low calcium | 3 | 0.17 | 2.2 | 13. | 1.2 | 0.084 |
| | 10 | 0.18 | 0.67 | 3.7 | 1.7 | 0.12 |
| Mean | | 0.18 | 1.4 | 8.4 | 1.4 | 0.10 |
| Normal calcium | 5 | 0.20 | 1.2 | 6.2 | 1.2 | 0.10 |
| | 8 | 0.15 | 1.0 | 6.7 | 1.3 | 0.083 |
| Mean | | 0.18 | 1.1 | 6.4 | 1.2 | 0.092 |
| High calcium | 4 | 0.10 | 5.1 | 51. | 1.3 | 0.072 |
| | 6 | 0.16 | 1.2 | 7.3 | 1.2 | 0.14 |
| Mean | | 0.13 | 3.2 | 29. | 1.2 | 0.11 |
| High strontium | 7 | 0.24 | 0.54 | 2.2 | 1.1 | 0.11 |
| | 9 | 0.18 | 0.77 | 4.3 | 1.5 | 0.13 |
| Mean | | 0.21 | 0.66 | 3.2 | 1.3 | 0.12 |

[a] From Comar and Wasserman, *A Symposium on Radioisotopes in the Biosphere*.[5]

the low observed ratios in the plasma and milk of the lactating cow and the high ratios in feces and urine.

Lough *et al.*[21] have extended these studies to the secretion of dietary $Sr^{90}$ and stable calcium in human milk. In four out of five normal, healthy women, the $Sr^{90}/Ca$ in milk ranged from 0.085 to 0.13 of the diet, with an average of 0.10. These values were essentially the same as reported for lactating animals. By assuming that the Sr/Ca ratio in the blood and milk was the same in the women studied as in the cow and goat, it was calculated that the Sr/Ca ratio in these subjects would be about 0.25 times that in the diet consumed.

The mechanism of calcium and strontium secretion from blood to milk was studied in goats by Twardock and Comar,[22] by using $Sr^{85}$, $Sr^{90}$, and $Ca^{45}$ in various combinations with continuous, long-term intravenous infusions as a means of maintaining desired blood levels of stable and radioisotopes of the two elements. Calcium and strontium secretion from blood to milk was the same whether the isotopes were given orally or intravenously; the same was true for radiostrontium excretion by the kidney. Infusions of stable Ca gave high blood-plasma levels and caused marked decreases in blood and milk levels of ingested $Ca^{45}$ and $Sr^{85}$ and reduced Ca/Sr discrimination. These results were attributed to flooding the renal tubules with stable Ca. The relative movement of calcium and strontium from blood to milk was unaffected by infusions. After injection into the mammary gland via the teat canal, equal amounts of radiocalcium and radiostrontium moved into the blood stream which indicated the existence of a unidirectional mechanism for preferential movement of calcium over strontium from blood to milk.

## References


1. G. C. H. Bauer, A. Carlsson, and B. Lindquist, Metabolism and Homeostatic Function of Bone, in *Mineral Metabolism*, Vol. I, pp. 609-676, Academic Press Inc., New York, 1961.
2. F. C. McLean and A. B. Hastings, Clinical Estimation and Significance of Calcium-ion Concentrations in the Blood, *Am. J. Med. Sci.*, **189**: 601-612 (1935).
3. W. F. Neuman and M. W. Neuman, *The Chemical Dynamics of Bone Mineral*, University of Chicago Press, Chicago, 1958.
4. J. Sendroy, Jr. and A. B. Hastings, Studies of the Solubility of Calcium Salts. II. The Solubility of Tertiary Calcium Phosphate in Salt Solutions and Biological Fluids, *J. Biol. Chem.*, **71**: 783-796 (1927).
5. C. L. Comar and R. H. Wasserman, Radioisotope Absorption and Methods of Elimination: Differential Behavior of Substances in Metabolic Pathways, in *A Symposium on Radioisotope in the Biosphere*, pp. 526-540, University of Minnesota, Minneapolis, 1960.
6. R. H. Wasserman, C. L. Comar, M. M. Nold, and F. W. Lengemann, Placental Transfer of Calcium and Strontium in the Rat and Rabbit, *Am. J. Physiol.*, **189**: 91-97 (1957).

7. D. Schachter and S. M. Rosen, Active Transport of $Ca^{45}$ by the Small Intestine and Its Dependence on Vitamin D, *Am. J. Physiol.*, **196**: 357-362 (1959).
8. D. Schachter, E. B. Dowdle, and H. Schenker, Accumulation of $Ca^{45}$ by Slices of the Small Intestine, *Am. J. Physiol.*, **198**: 275-279 (1960).
9. D. Schachter, Vitamin D and the Active Transport of Calcium by the Small Intestine, in *Conf. on the Transfer of Calcium and Strontium across Biological Membranes* (in press), Academic Press, New York, 1963.
10. H. Rasmussen, The Influence of Parathyroid Function upon the Transport of Calcium in Isolated Sacs of Rat Small Intestine, *Endocrinology*, **65**: 517-519 (1959).
11. R. H. Wasserman, Metabolic Basis of Calcium and Strontium Discrimination: Studies with Surviving Intestinal Segments, *Proc. Soc. Exptl. Biol. Med.*, **104**: 92-95 (1960).
12. G. E. Harrison, E. Lumsden, W. H. A. Raymond, and A. Sutton, On the Mechanisms of Skeletal Fixation of Strontium. Part I, *Arch. Biochem. Biophys.*, **80**: 97-105 (1959).
13. R. H. Wasserman, F. A. Kallfelz, and C. L. Comar, Active Transport of Calcium by Rat Duodenum in vivo, *Science*, **133**: 883-884 (1961).
14. F. W. Lengemann and C. L. Comar, Distribution of Absorbed Strontium-85 and Calcium-45 as Influenced by Lactose, *Am. J. Physiol.*, **200**: 1051-1054 (1961).
15. R. Nicolaysen and N. Eeg-Larsen, The Biochemistry and Physiology of Vitamin D, *Vitamins and Hormones*, **11**: 29-60 (1953).
16. R. Nicolaysen, The Absorption of Calcium as a Function of the Body Saturation with Calcium, *Acta Physiol. Scand.*, **5**: 200-211 (1943).
17. R. Nicolaysen, N. Eeg-Larsen, and O. J. Malm, Physiology of Calcium Metabolism, *Physiol. Rev.*, **33**: 424-444 (1953).
18. R. Haavaldsen, K. M. Egnund, and R. Nicolaysen, Studies in Calcium Metabolism in Rats. II. The Interaction of Vitamin D and the Endogenous Factor, *Acta Physiol. Scand.*, **36**: 108-113 (1956).
19. A. Carlsson, Experiments with Radiocalcium on the Interrelationships Between Vitamin D and Dietary Calcium and Phosphorus, *Acta Pharmacol. Toxicol.*, **9**: 32-40 (1953).
20. H. Spencer, D. Laszlo, and M. Brothers, $Strontium^{85}$ and $Calcium^{45}$ Metabolism in Man, *J. Clin. Invest.*, **36**: 680-688 (1957).
21. S. A. Lough, G. H. Hamada, and C. L. Comar, Secretion of Dietary Strontium 90 and Calcium in Human Milk, *Proc. Soc. Exptl. Biol. Med.*, **104**: 194-198 (1960).
22. A. R. Twardock and C. L. Comar, Calcium and Strontium Secretion from Blood to Milk, *Am. J. Physiol.*, **201**: 645-650 (1961).
23. C. L. Comar, I. B. Whitney, and F. W. Lengemann, Comparative Utilization of Dietary $Sr^{90}$ and Calcium by Developing Rat Fetus and Growing Rat, *Proc. Soc. Exptl. Biol. Med.*, **88**: 232-236 (1955).

# Chapter 9

# *Homeostasis in Mineral Metabolism*

Until recent years it has been customary to regard the kidneys as functioning in an automatic fashion, with built-in mechanisms which respond to alterations in the composition of the blood plasma so as to conserve elements in short supply and to eliminate those in excess. More recently, however, it has become apparent that the ability of the kidneys to act automatically is limited and that increasing importance must be ascribed to central and hormonal control. We have advanced the hypothesis[1] that the major cations of the blood plasma—$Na^+$, $K^+$, $Ca^{2+}$, $Mg^{2+}$, and $H^+$—are subject to such extrarenal control.

For the major anions of the blood plasma—$Cl^-$, $HCO_3^-$, $H_2PO_4^-/HPO_4^{2-}$, and $SO_4^{2-}$—the situation is quite different. As against the generalization that the cations of the blood are subject to central or hormonal regulation, the contrary statement may be made for the anions, i.e., that the control of none has been shown to depend upon information transmitted from a higher center and that the existence of a self-regulatory system has been found for none of these anions.

## 1. Homeostasis of $Ca^{2+}$ in the Fluids of the Body

Higher organisms possess systems which comprise mineralization and resorption of bone in addition to exchange of the ions of the bone mineral with the fluids of the body. Closely allied with these physiological activities is the homeostatic control of the calcium ion concentration in the blood plasma and in the other extracellular fluids. In the functioning of these systems, the organism controls absorption of calcium and phosphate and their excretion, making use of the gastrointestinal system and the kidneys, both for conservation of needed elements and for elimination of any excess. Operation of such complex and interrelated functions requires integration of regulatory mechanisms. Of the controls, the most important, in so far as the metabolism of calcium and of phosphate is concerned, are vitamin D (both exogenous and endogenous) and the parathyroid hormone. So closely interrelated are the functional activities of the parathyroid hormone and vitamin D that it will serve our purposes to consider them both as component parts of an integrated system which exercises control over the mechanisms required for proper functioning of the organism as a whole.

It was suggested very early that the vitamin and the hormone might act by influencing each other; this, however, has never been demonstrated, and it is now assumed that each makes its own contribution to the regulatory processes. Thus, the parathyroid glands monitor the calcium ion concentration in the plasma and apply the necessary corrections, mainly by influencing the mobilization of calcium from the skeleton. The parathyroid hormone also exercises control over the excretion of phosphate by the kidneys and aids in influencing the absorption of calcium in the gastrointestinal tract. Vitamin D, on the other hand, is the major influence upon the absorption of calcium; it controls the mineralization of bone by assuring an adequate supply of the necessary materials. Vitamin D also exerts an effect upon mobilization of mineral from bone, complementing the action of the parathyroid hormone in this respect. An oversimplified version of current views is that vitamin D influences mobilization in proportion to its formation and intake and that the parathyroid glands are called upon to balance the interplay between dietary intake, absorption, urinary and fecal excretion of calcium, and the regulatory effect of vitamin D. According to this view, the parathyroid glands play the decisive role in maintaining the $Ca^{2+}$ concentration in a steady state.

## 2. Cybernetic Aspects of Homeostasis of $Ca^{2+}$

The $Ca^{2+}$ concentration in the blood is itself the stimulus to which the parathyroid glands respond; the condition being regulated activates the regulatory mechanism, feeding back information about the output to an earlier stage so as to influence its action and thereby control the output. This is a clear-cut example of *negative feedback*, in terms of *cybernetic* or *control system theory*, and will be so considered here.

The term *cybernetics*, as now used, includes the whole field of control and information theory, both in the machine and in biological systems. The word itself is derived from a Greek word, meaning "steersman." *Information theory*, in the biological sense, refers to the type of information transmitted by humoral or nervous pathways capable of eliciting a physiological response. In such a system, the deviation from the normal is called the *error*. The term *negative feedback* refers to reduction of the error and may operate, in the case of $Ca^{2+}$ concentrations, to correct a concentration that is either too high or too low. A *closed cycle system*, such as that under consideration, is also called a *servo system*. There are many servo systems in living organisms.

A servo system characteristically includes three parts: (1) an *error sensing device;* (2) a *feedback transducer*, which translates the *error signal* from one form to another, e.g., a chemical signal to a change in the rate of output of an

internal secretion; and (3) a *control signal transducer,* which brings about a further change in the nature of the signal in such a fashion as to affect the *output.* In biological systems, two of these functions may be included within a single physiological unit.

In many approaches to the elucidation of homeostatic control in biological systems, the analogy has been to the common thermostatic control of temperature. In such a system there are only two possible settings for corrective action—on and off. This is called *directed correction.* One could question whether such a crude type of control ever exists in a biological system. A second type of control is called *proportional control;* the action of the control mechanism

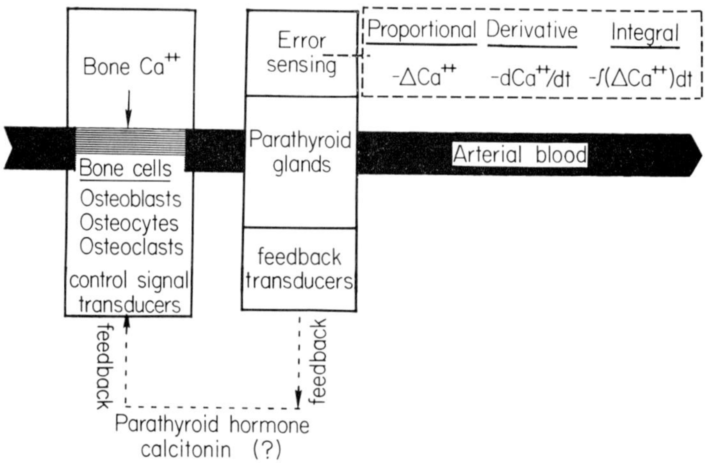


FIG. 9.1. Diagram of control system regulating $Ca^{2+}$ concentration in the blood.

is proportional to the deviation from the norm, i.e., proportional to the error. This is probably the simplest type of control found in biological systems.

A third type of control makes use of proportional control but adds *derivative control* in which the action is proportional to the rate of change of the error and is useful in preventing overshoot. This is used in gunfire control to anticipate the location of the target on the basis of its present velocity. Derivative control is postulated for biological systems in which overshoot, possibly leading to oscillation, would be an unfavorable consequence of regulatory action.

A fourth type of control includes proportional and derivative control but adds *integral control,* sensitive to the accumulation of error with time. Some degree of integral control is doubtless present in most of the biological systems required for physiological regulation.

Figure 9.1 is a diagram, in terms of control theory, of the control mecha-

nism responding to a deviation in the $Ca^{2+}$ concentration in the blood as mediated through the parathyroid gland. This diagram illustrates proportional, derivative, and integral control, all with reference to the $Ca^{2+}$ concentration in the blood plasma reaching the parathyroid glands. The glands act first as error-sensing devices, and then as feedback transducers by translating the stimulus received from the $Ca^{2+}$ concentration into variations in output of the parathyroid hormone(s). The hormone is then carried in the blood to the bones, where it influences the mobilization of $Ca^{2+}$ by cellular action. The cells participating in this action should be regarded as control-signal transducers, since they respond to control by hormones, and act in some other manner, not as yet understood, as discussed below. Such a response occurs most clearly from osteoclasts—it is commonly agreed that the parathyroids regulate osteoclastic resorption of bone. It appears, in part from the observations of Heller *et al.*,[2] that osteocytes also respond to parathyroid hormone. There is some evidence from the results of *in vitro* studies with glycine-$C^{14}$ for participation of osteoblasts; collagen synthesis is impaired in cultures of bone in the presence of parathyroid hormone.[3]

## 3. The Parathyroid Hormone(s)

In recent years it has been commonly believed that the parathyroid glands produce one internal secretion (i.e., the parathyroid hormone) which acts in one direction only to control the release of $Ca^{2+}$ from bone to blood. The idea of a second hormone, responding to the stimulus of varying concentrations of phosphate ions in the plasma, is now on the decline, although it is still held in some quarters. In its place, however, there is now postulated another hormone, called *calcitonin,* that is responsive to hypercalcemia and acts to lower the plasma calcium level. If the secretion of calcitonin is confirmed (the evidence for it is reviewed below), some revision of terminology will be necessary since one will no longer be able to speak of *the* parathyroid hormone.

Rasmussen and Craig[4] have isolated and characterized a single protein with a molecular weight of approximately 9500, which (1) is homogeneous by countercurrent distribution, paper and column chromatography, and ultracentrifugation, (2) contains no cystine, and (3) possesses a single end-terminal amino acid, alanine. The amino acid composition of the pure bovine hormone is known. The present evidence indicates that this hormone consists of a single polypeptide chain which can be partially hydrolyzed without complete loss of biological activity. The pure preparation has both calcium-mobilizing and phosphaturic activity; demonstration of both these properties in a single purified substance is evidence against another hormone with the specific function of regulating phosphate excretion. Calcitoninlike activity has

not been demonstrated for this purified hormone, which has been found present in crude parathyroid extract.[5]

## 4. Calcitonin

Copp and his collaborators[6,7] have reviewed the evidence for a second hormone from the parathyroid glands—which they have named *calcitonin.* In fasting anesthetized dogs, these investigators perfused isolated thyroid-parathyroid glands with high-calcium blood (13 mg/100 cm$^3$); the result was an immediate fall in systemic blood calcium, even in the presence of intact thyroid and parathyroid glands on the other side of the neck. The fall of the plasma calcium resulting from parathyroid hypercalcemia was greater than that following total surgical thyroparathyroidectomy. The difference in response was still more striking after a period during which production of parathyroid hormone had been stimulated by perfusing the parathyroids with low-calcium blood. In such animals, parathyroidectomy was followed by a sustained rise in plasma calcium, while high-calcium perfusion of the glands resulted in the usual prompt fall. The evidence was interpreted to indicate that hypercalcemia causes the release of a hypocalcemic factor from the parathyroids that lowers the blood calcium and antagonizes the effect of the hypercalcemic factor, until now known as the parathyroid hormone. Such a hypocalcemic factor has not been isolated, but its activity has been demonstrated in hypercalcemic perfusates of the parathyroids, in commercial parathyroid extracts, and in preparations free of hypercalcemic activity. An analogy, if it is needed, is that the islets of the pancreas produce two hormones—insulin and glucagon—both of which are involved in the regulation of the level of blood glucose.

In the presence of the parathyroid glands, the prevention of overshoot following induced hypocalcemia could be attributed to the influence of derivative control. In the absence of the parathyroids, the control is changed from a closed-cycle to an open-cycle system; i.e., the error sensing device and the feedback transducer are eliminated, and there is no feedback. That there is still a crude regulatory mechanism, which transfers $Ca^{2+}$ from bone to blood and maintains the $Ca^{2+}$ concentration in the plasma at a level usually compatible with life, is a feature of the dual mechanism to be described below.

## 5. The Dual Mechanism

McLean and Urist[8] have postulated a dual mechanism in homeostatic regulation of the $Ca^{2+}$ concentration in the blood plasma. One part, dependent upon parathyroid activity, has been described above in terms of cybernetic theory and will be further considered in terms of the mode of action of the parathyroid hormone on bone.

The other part of the dual mechanism, operative in the absence of the parathyroid glands, remains to be accounted for. Beginning with the experiments of Hastings and Huggins,[9] it has been repeatedly demonstrated that a rapid exchange of $Ca^{2+}$ ions continues between bone and blood after removal of the parathyroid glands. This exchange results in the maintenance of a relatively constant, but low, level of $Ca^{2+}$ in the blood plasma. McLean and Urist attributed this to passive physicochemical equilibrium between the body fluids and the labile fraction of the bone mineral, not requiring a feedback mechanism.

Another way of accounting for this is to assume that transfer of calcium from bone to blood is, as postulated by Neuman and Dowse,[10] at all times under the influence of the $H^+$ ions produced in bone as a product of the glycolytic cycle and that the *rate* of their production is under parathyroid control. One need only assume that the osteocytes do, in fact, produce $H^+$ and that these hydrogen ions aid in the solubilization of bone mineral and in its liberation into the body fluids. There is both chemical and histochemical evidence in favor of this view. Neuman *et al.*[11] and others have shown that lactic and pyruvic acid production in bone is influenced by parathyroid activity. Heller *et al.*[2] have found zones of increased reactivity of bone mineral around the lacunae housing osteocytes after administration of parathyroid extract.

In any event, it is clear that the body fluids have access to the labile or reactive fraction of the bone mineral, even in the absence of the parathyroid glands. It has also been demonstrated, initially by Woods and Armstrong,[12] that administration of parathyroid extract to rats with $Ca^{45}$ incorporated within the stable fraction of their bone mineral is capable of releasing the isotope into the blood stream—this is evidence that the parathyroid glands control osteoclastic resorption of the stable mineral.

## 6. Mode of Action of the Parathyroid Hormone on Bone

It is no longer necessary to defend the thesis that the parathyroid hormone acts directly on bone. It should be stated, however, that there is little or no evidence for a direct effect of the hormone on the solubilization of the bone mineral or on depolymerization and solubilization of the organic matrix of bone. There is every reason to believe that the effects of the hormone on both the mineral and organic constituents of bone are mediated by cells. As assumed above, the three types of the cells of bones—osteoclasts, osteocytes, and osteoblasts—are capable of acting as control signal transducers in response to the stimulus of the parathyroid hormone circulating in the blood stream. It is equally probable that the hormone acts to control the *rates* of cellular activities and modulations which continue, at a lower rate, in its absence.

Neuman and his collaborators[10,11,13] have introduced a unifying concept

of parathyroid hormone action which concludes that all aspects of the action of the hormone can be explained in terms of its ability to enhance the transport of inorganic phosphate into cells with, in bone, an associated increase in acid production by glycolysis. To apply this postulate to the influence of the parathyroid hormone upon the transfer of exchangeable calcium from bone to blood, in terms of the localization of the labile pool of calcium in the new and incompletely calcified osteons, requires only emphasis on the glycolytic cycle in these cells and enhancement of the solubilization of bone mineral by the effects of the acid so produced. To further apply the postulate to osteoclastic resorption, which occurs at a different place and under different conditions in that it results in resorption of whole bone as well as solubilization of the bone mineral, requires an additional explanation. This has also been supplied by Neuman *et al.*,[14] who have developed a chemical view of osteoclasis based on studies with radioactive yttrium. They conclude that osteoclasts are required for the removal of bone and that this process depends upon both acid, or other solubilizers of bone mineral, and proteolytic enzymes.

Except for the evidence that a second hormone from the parathyroid glands, calcitonin, acts in the opposite direction from that of the known parathyroid hormone, there is no evidence bearing on the mode of action of calcitonin. In the absence of such information, Fig. 9.1 shows calcitonin acting through the same channels as does parathyroid hormone but with an opposing effect.

## 7. Use of Isotopes in the Study of Parathyroid Hormone

Elucidation of parathyroid function, and of the mode of action of the parathyroid hormone on bone, has not depended primarily upon the use of isotopic methods. Mention has been made above of the demonstration, by Woods and Armstrong,[12] of the ability of parathyroid activity to promote osteoclastic resorption of the stable bone mineral; their conclusions are based on the liberation of $Ca^{45}$ from this fraction of the mineral. Reference has also been made of the use, by Neuman *et al.*,[14] of radioactive yttrium, $Y^{90}$, in the study of osteoclasis.

Talmage *et al.*[15] have also shown that parathyroid extract releases $Sr^{85}$ and $P^{32}$ from bone; this effect was found only when the isotope had been in bone for a long period of time—and the conclusion was drawn that the isotope was released to the body fluids only by action of the parathyroids on dissolution of bone mineral in areas of bone in which the crystals were not readily available for equilibration with the fluid compartment.

Shetlar *et al.*[16] have used sulfate-$S^{35}$ as a tracer in an attempt to learn whether the sulfated mucopolysaccharides of bone are affected by parathyroid extract. It was found that administration of the extract caused increased release of sulfated mucopolysaccharides into the blood stream and it was suggested

that the major mucopolysaccharide so liberated was chondroitin sulfate. This is not, however, evidence that the hormone acts directly on the ground substance of bone; mediation by the cellular elements appears more probable. The same workers demonstrated that both $Ca^{45}$ and $S^{35}$ were liberated into the serum of rats, in a parallel fashion, when parathyroid extract was administered to animals previously injected with either of these isotopes at times selected in order to allow them to be incorporated into bone. Similar results have been reported by Bronner[17] after administration of the same isotopes ($Ca^{45}$ and $S^{35}$) to rats, followed by treatment of the animals with parathyroid extract.

Clark[18] studied the effect of agents influencing the removal of skeletal radiocalcium and radiophosphorus from intact and parathyroidectomized rats. He found that magnesium, citrate, vitamin D, and parathyroid extract appeared to remove calcium preferentially from bone in both series of animals, although in some instances there was a different pattern of response in the parathyroidectomized rats. The removal of the parathyroid glands seemed to result in a more rapid exchange of calcium and phosphorus between bone and tissue fluids than that in intact rats. The actions of vitamin D and of parathyroid extract on calcium and phosphorus metabolism appeared similar in their effects on urinary excretion of these elements.

It is only recently that the relationship between the parathyroids and absorption of calcium has been studied[19,20]; it is now generally accepted that the parathyroid hormone promotes absorption of calcium in the intestine. Cramer *et al.*[21] have concluded from the results of balance studies that parathyroid extract causes increased absorption of dietary calcium from the gastrointestinal tract. They also reported that an effect of parathyroidectomy was demonstrated in young rats when carrier calcium was added to the administered $Ca^{45}$. In more recent work employing tracer methods, there seems to be adequate evidence for an effect of parathyroid hormone on intestinal absorption of calcium. Less work has been done on the effect of the parathyroids on strontium absorption, but Bacon *et al.*[22] reported that parathyroid extract increases the urinary excretion of Sr as well as of Ca. In the absence of evidence to the contrary, it is probably safe to assume that if parathyroid hormone has a positive effect upon the intestinal absorption of calcium the same will be found true for strontium.

## 8. Vitamin D

Vitamin D, whether endogenous or exogenous, may be regarded as essential to the total system that includes calcium metabolism and the mineralization of bone; its effects are made manifest in a variety of ways.

The antirachitic effect of vitamin D in prophylactic doses is effected mainly by its influence upon intestinal absorption of calcium. In very large doses (50,000 I.U. or more daily), vitamin D exerts a marked calcemic effect, which results in an increased concentration of calcium in the blood plasma—an effect similar to that produced by parathyroid extract. This subtoxic action of the vitamin and of other chemically related sterols is used to raise the level of serum calcium in hypoparathyroidism. In even larger doses, the effects are manifestly toxic and result in hypervitaminosis D, a condition which is associated with resorption of bone comparable to that induced by large doses of parathyroid extract.

What appears to have been the first report of the use of radioisotopes in demonstrating the effect of vitamin D upon the absorption of calcium and strontium from the digestive tract was published in 1945 by Greenberg.[23] At the time Greenberg's work was done, $Ca^{45}$ and $Sr^{89}$ were only available from bombardment of the metals with deuterons in a cyclotron, and the isotopes were administered in solution as their lactates. Addition of vitamin D to the diets promoted absorption of both Ca and Sr in rachitic rats, whether the isotope was administered orally or by injection. No calculations of the relative effectiveness on calcium and strontium absorption were reported. Greenberg also concluded that vitamin D exerts a direct effect on the mineralization of bone in rachitic animals. This possibility is now largely discounted, and in its place there is the view that the direct effect of the vitamin on bone leads to mobilization of calcium rather than to its deposition.

There can be no doubt that vitamin D, in large doses, promotes the resorption of bone in a manner resembling the effects of parathyroid hormone. This is the calcemic effect of vitamin D, reviewed at length by McLean.[24] Nicolaysen and Eeg-Larsen[25] have also emphasized a distinction between the effects of physiological and pharmacodynamic doses of the vitamin, stating that in the bones the net result of the pharmacodynamic effect is the opposite of the physiological effect.

On the other hand, Carlsson[26,27] has found that even in small doses vitamin D favors the removal of lime salts from the bones and has regarded the other effects of the vitamin (increased serum calcium, increased growth, and calcification) as secondary effects. Lindquist[28] observed that a dose of 15 I.U. of vitamin D administered to rats produced a maximal effect on the absorption of calcium from the digestive tract, while doses up to 50,000 I.U. resulted in an accentuation of the direct physiological effect on bone. By making use of Lindquist's data, Bauer *et al.*[29] concluded that bone resorption is an active process which requires the presence of vitamin D for its normal homeostatic function and that the earlier concept that the antirachitic and calcemic effects are caused by different mechanisms could no longer be maintained; they

rather represent different degrees of action. Nicolaysen and Eeg-Larsen[30] repeated the experiments of Carlsson with similar results, and came to similar conclusions, as have Causeret and Hugot[31] and Neuman and Neuman.[32]

The basic experiments of Carlsson and his collaborators were performed under conditions of severe restriction of the animal's intake of calcium. The evidence they bring forward indicates that, under such conditions, there is a transfer of calcium from one part of the skeleton to another, apparently to meet local needs. When calcium is added to the diet, this effect is added to the influence of vitamin D on the absorption of calcium by the intestines. The result is that there are both a redistribution of the calcium in the skeleton and a net increase in the total amount present. If such a redistribution occurs, it may be regarded as a recycling of calcium within the skeleton that is promoted by the action of vitamin D. It thus appears that the skeleton can make its needs known to the absorptive mechanism, possibly through the medium of the endogenous factor of Nicolaysen[33] and also through an unknown pathway, to other locations in the skeleton. Both mechanisms require vitamin D for their activation. It will be noted also that the recycling of calcium within the skeleton appears to be carried out by influences upon the resorption rather than upon accretion of the bone mineral. The stimulus for recycling, however, as well as for increased intestinal absorption may be regarded as originating within the parts of the skeleton in need of additional mineral.

## References


1. F. C. McLean, Introduction: Homeostasis in Mineral Metabolism, in *Mineral Metabolism*, Vol. I, pp. 1-10, Academic Press Inc., New York, 1960.
2. M. Heller, F. C. McLean, and W. Bloom, Cellular Transformations in Mammalian Bones Induced by Parathyroid Extract, *Am. J. Anat.*, **87**: 315-348 (1950).
3. P. J. Gaillard, Parathyroid and Bone in Tissue Culture, in *The Parathyroids*, pp. 20-48, Charles C Thomas, Springfield, Illinois, 1961.
4. H. Rasmussen and L. C. Craig, Isolation and Characterization of Bovine Parathyroid Hormone, *J. Biol. Chem.*, **236**: 759-764 (1961).
5. D. H. Copp and E. C. Cameron, Demonstration of a Hypocalcemic Factor (Calcitonin) in Commercial Parathyroid Extract, *Science*, **134**: 2038 only (1961).
6. D. H. Copp and A. G. F. Davidson, Direct Humoral Control of Parathyroid Function in the Dog, *Proc. Soc. Exptl. Biol. Med.*, **107**: 342-344 (1961).
7. D. H. Copp, E. C. Cameron, B. A. Cheney, A. G. F. Davidson, and K. G. Henze, Evidence for Calcitonin—A New Hormone from the Parathyroid that Lowers Blood Calcium, *Endocrinology*, **70**: 638-649 (1962).
8. F. C. McLean and M. R. Urist, *Bone: An Introduction to the Physiology of Skeletal Tissue*, 1st ed., University of Chicago Press, Chicago, 1955.
9. A. B. Hastings and C. B. Huggins, Experimental Hypocalcemia, *Proc. Soc. Exptl. Biol. Med.*, **30**: 458-459 (1933).
10. W. F. Neuman and C. M. Dowse, Possible Fundamental Action of Parathyroid Hormone in Bone, in *The Parathyroids*, pp. 310-330, Charles C Thomas, Springfield, Illinois, 1961.

11. W. F. Neuman, H. Firschein, P. S. Chen, Jr., B. J. Mulryan, and V. DiStefano, On the Mechanism of Action of Parathormone, *J. Am. Chem. Soc.*, **78**: 3863-3864 (1956).
12. K. R. Woods and W. D. Armstrong, Action of Parathyroid Extracts on Stable Bone Mineral Using Radiocalcium as Tracer, *Proc. Soc. Exptl. Biol. Med.*, **91**: 255-258 (1956).
13. A. R. Terepka, C. M. Dowse, and W. F. Neuman, *Unifying Concepts of Parathyroid Hormone Action*, USAEC Report UR-577, pp. 1-28, Univ. Rochester Atomic Energy Project, September 1960.
14. W. F. Neuman, B. J. Mulryan, and G. R. Martin, A Chemical View of Osteoclasis Based on Studies with Yttrium, *Clin. Orthopaed.*, **17**: 124-134 (1960).
15. R. V. Talmage, L. T. Wimer, and R. J. Toft, Additional Evidence in Support of McLean's Feedback Mechanism of Parathyroid Action on Bone, *Clin. Orthopaed.*, **17**: 195-205 (1960).
16. M. R. Shetlar, R. H. Bradford, W. Joel, and R. P. Howard, Effects of Parathyroid Extract on Glycoprotein and Mucopolysaccharide Components of Serum and Tissue, in *The Parathyroids*, pp. 144-157, Charles C Thomas, Springfield, Illinois, 1961.
17. F. Bronner, Parathyroid Effects on Sulfate Metabolism: Interrelationships with Calcium, in *The Parathyroids*, pp. 123-143, Charles C Thomas, Springfield, Illinois, 1961.
18. I. Clark, Agents Effecting the Removal of Skeletal Radiocalcium and Radiophosphorus from Intact and Parathyroidectomized Rats, in *The Parathyroids*, pp. 183-202, Charles C Thomas, Springfield, Illinois, 1961.
19. R. V. Talmage and R. J. Toft, The Problem of the Control of Parathyroid Secretion, in *The Parathyroids*, pp. 224-242, Charles C Thomas, Springfield, Illinois, 1961.
20. H. Rasmussen, Parathyroid Hormone. Nature and Mechanism of Action, *Am. J. Med.*, **30**: 112-128 (1961).
21. C. F. Cramer, A. P. Suiker, and D. H. Copp, Parathyroid Influence on Calcium and Phosphorus Absorption by the Gut, in *The Parathyroids*, pp. 158-166, Charles C Thomas, Springfield, Illinois, 1961.
22. J. A. Bacon, H. Patrick, and S. L. Hansard, Some Effects of Parathyroid Extract and Cortisone on Metabolism of Strontium and Calcium, *Proc. Soc. Exptl. Biol. Med.*, **93**: 349-351 (1956).
23. D. M. Greenberg, Studies in Mineral Metabolism with the Aid of Artificial Radioactive Isotopes. VIII. Tracer Experiments with Radioactive Calcium and Strontium on the Mechanism of Vitamin D Action in Rachitic Rats, *J. Biol. Chem.*, **157**: 99-104 (1945).
24. F. C. McLean, Activated Sterols in the Treatment of Parathyroid Insufficiency, in *Glandular Physiology and Therapy*, pp. 461-493, American Medical Association, Chicago, Illinois, 1942.
25. R. Nicolaysen and N. Eeg-Larsen, The Biochemistry and Physiology of Vitamin D, *Vitamins and Hormones*, **11**: 29-60 (1953).
26. A. Carlsson, Tracer Experiments on the Effect of Vitamin D on the Skeletal Metabolism of Calcium and Phosphorus, *Acta Physiol. Scand.*, **26**: 212-220 (1952).
27. A. Carlsson, Experiments with Radiocalcium on the Interrelationships Between Vitamin D and Dietary Calcium and Phosphorus, *Acta Pharmacol. Toxicol.*, **9**: 32-40 (1953).
28. B. Lindquist, Effect of Vitamin D on the Metabolism of Radiocalcium in Rachitic Rats, *Acta Paed.*, **41**: Suppl. 86 (1952).

29. G. C. H. Bauer, A. Carlsson, and B. Lindquist, Evaluation of Accretion, Resorption, and Exchange Reactions in the Skeleton, *Kungl. Fysiol. Sällskapets Lund Förhandl.*, **25**: 1-16 (1955).
30. R. Nicolaysen and N. Eeg-Larsen, The Mode of Action of Vitamin D, in *Ciba Foundation Symposium on Bone Structure and Metabolism*, pp. 175-186, Little, Brown & Co., Boston, 1956.
31. J. Causeret and D. Hugot, Effet de la Vitamine D sur la Calcémie du Jeune Rat Carencé en Calcium et sur le Développement de Diverses Parties de Son Squelette, *Bull. Soc. Chim. Biol.*, **41**: 843-853 (1959).
32. W. F. Neuman and M. W. Neuman, *The Chemical Dynamics of Bone Mineral*, The University of Chicago Press, Chicago, 1958.
33. R. Nicolaysen, The Absorption of Calcium as a Function of the Body Saturation with Calcium, *Acta Physiol. Scand.*, **5**: 200-211 (1943).

# Chapter 10

# *Application of Tracer Techniques to Bone*

## 1. Introductory Statement

Tracer techniques, in the broadest sense, have been applied to the study of bone for more than two centuries. By using metal markers, Duhamel[1] demonstrated that bones grow by apposition and that interstitial growth in length does not occur. These techniques were later refined by the use of roentgenograms for *in vivo* localization of metal markers or of other landmarks, such as growth-arrest lines, exostoses, or striae caused by lead, bismuth, or phosphorus poisoning. The use of chemical markers dates from Belchier,[2,3] who in 1736 demonstrated vital staining of newly forming bone with madder and thus made possible a detailed description of bone remodeling in experimental animals. Belchier's technique was later supplanted by the use of alizarin. In 1926 Aub and his collaborators[4] used lead in intravital staining of bone, and Vincent[5] demonstrated the distribution of lead in osteons during osteogenesis and in those in which calcification was complete. Milch *et al.*[6] in 1958 made a further step forward with chemical marking by the introduction of the tetracyclines, which act as fluorescent dyes when administered in the living subject.

The use of $P^{32}$ by Chiewitz and Hevesy,[7] and of $Ca^{45}$ by Greenberg,[8] has already been noted; since then, advance in the employment of radioactive tracers in the study of bone has been rapid. The physiological processes most amenable to study by means of radioisotopes are: (1) the movement of elements and compounds, including the phenomena of uptake, turnover, exchange and transfer; and (2) the metabolic pathways and concentration of radioelements, their ionic groupings, or the compounds that may be labeled with them. All of these physiological activities may be studied qualitatively and quantitatively by the use of radioactive tracers. The methods fall roughly into autoradiography, radioassay, and external scanning or whole-body counting.

The application of any particular method is largely governed by the nature of the problem, so that a combination of methods is often necessary to correlate the results from one method with those of another. For example, autoradiography and distribution studies will give information on deposition and localization; patterns of distribution, radioassay, chromatography, electrophoresis, and *in vitro* methods can furnish results on the nature of the deposi-

tion; chemical assay, studies of metabolites, labeled molecules, and multiple tracer studies can furnish additional information on metabolism of the radionuclide.

## 2. Autoradiography

One of the most fruitful and most frequently used tools of research in the study of bones and teeth is that of autoradiography. For theories, methods, interpretation, and general information, the reader is referred to the comprehensive work of Boyd[9] and the earlier publication of Yagoda.[10] Methods applicable to bone are described by Comar.[11] An excellent account of techniques and applications is presented in a conference on autoradiography with an emphasis on automation in quantitative evaluation.[12]

### 2.1. Qualitative Autoradiography

The principles of autoradiography are applicable to the deposition and localization of the radioisotope, for visualization of patterns of distribution, and to bone growth and development. The sample containing radioactive material is placed in contact with a photographic emulsion, and, after development, an image is produced which provides visualization of the location of the radioactivity in the sample. On a gross scale, a whole bone or a small animal may be cut with a saw, and the cut surface placed in contact with film; thus the distribution of the radioactivity may be viewed in either sagittal or cross section.

The technique for serial autoradiographs for recording patterns of distribution has been reported by Engström *et al.*[13] After administration of $Sr^{90}$, rats were sacrificed at various intervals from 5 minutes to 128 days. The animals were frozen at very low temperatures and were then cut longitudinally. The cut surface was placed on industrial X-ray film and exposed for 40 minutes. Table 10.1 lists the areas of radioactivity concentration in organs and different parts of the skeleton at various intervals of time.

A method which is coming into wide use for the localization and distribution of a radionuclide has been developed by Ullberg[14] and is a technique in which the mouse or rat is frozen, embedded in ice, and sectioned with a heavy-duty microtome. The sagittal sections are then placed on high-resolution emulsion and exposed for appropriate lengths of time (Fig. 6.4). This method is especially useful for labeled water-soluble substances, for short-lived radionuclides such as $F^{18}$, and for tritiated estrogens where routine fixatives are to be avoided.

Autoradiographs made from thin ground sections of cortical bone are easiest to interpret. The early work of Amprino[15] which combined the micro-

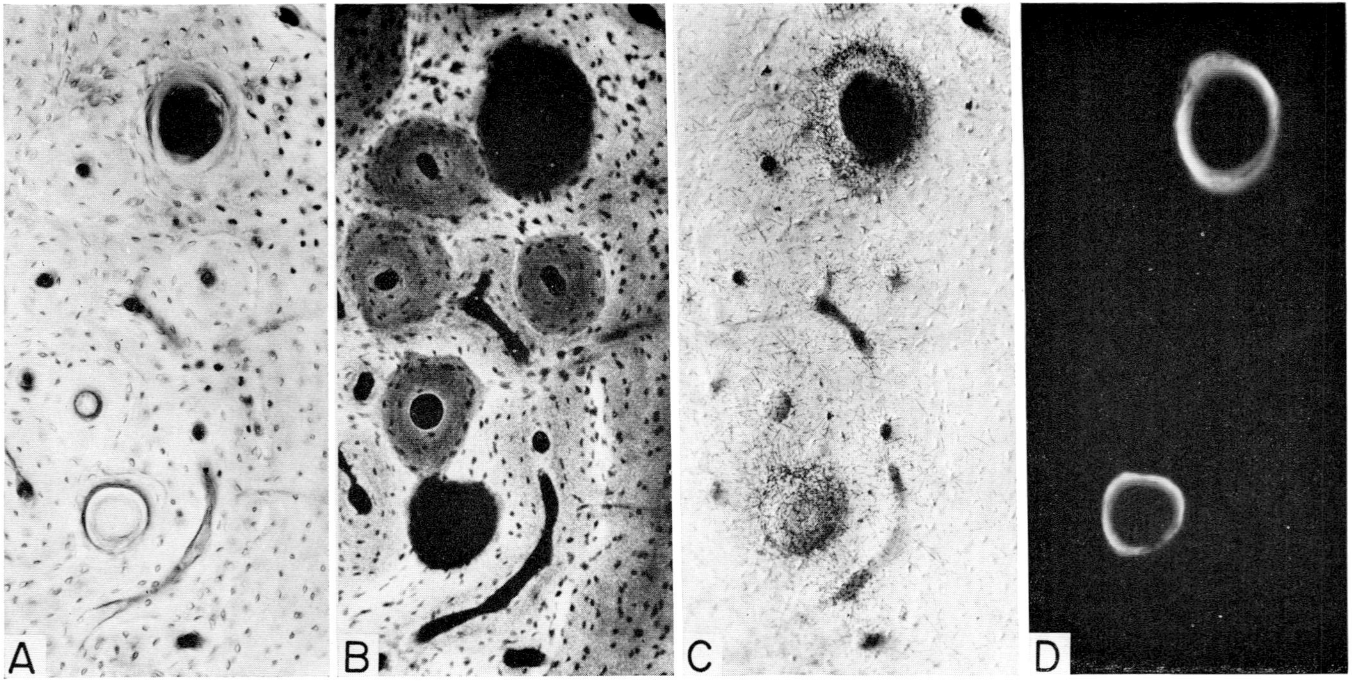


FIG. 10.1. Four illustrations of the same section of compact bone from the radius of a dog. The bone is from an 8-month-old beagle, given 35 mg/kg tetracycline for 4 days and 40 μc $Ra^{226}$ for 2 days before sacrifice, both intravenously. The bone was embedded in methyl methacrylate and the section, 100 μ thick, was cut with a high-speed rotary saw, without further treatment. A, Photomicrograph of unstained section; B, microradiograph, made with 9-kv X-rays on Eastman Kodak 649-0 spectroscopic plate; C, alpha-track autoradiograph, made with stripping film, Kodak Limited AR-10, exposed for 14 days; D, fluorescent image with ultraviolet light, photographed on Kodak Panatomic X film. (All images × 154.) (From R. E. Rowland, Argonne National Laboratory.)

TABLE 10.1

DISTRIBUTION OF $Sr^{90}$ IN AUTORADIOGRAMS OF RATS SACRIFICED AFTER DIFFERENT INTERVALS[a]

| Organs | Interval after injection of $Sr^{90}$ (10 μC) intraperitoneally | | | |
|---|---|---|---|---|
| | 5 minutes | 15 minutes | 1 hour | 6 hours |
| Viscera | Diffuse distribution. Highest density in abdominal cavity. Increased in lungs but not in heart | Decreased density in abdominal cavity and lungs | Increased density practically disappeared in abdominal cavity and completely disappeared in lungs | Dense spot in lower part of abdominal cavity |
| Skull | Diffuse distribution, low density | Density increased | Density further increased. Details clearly seen | Density reaches maximum |
| Incisors | Same density as in skull but a little increased at basis | Same density as in skull but a little increased at basis | Same density as in skull but a little increased at base | Very high density of incisors |
| Spine | Low density but details can be seen | Density increased | Density further increased. Details clearly seen | Density reaches maximum |
| Tail | No increased density | Density in tail can be observed | As above | As above |
| Ribs, Pubic bone, Tracheal rings | Low density Details can be seen | Density increased | Density further increased | As above |

[a] From Engström, *et al.*, *Bone and Radiostrontium*.[13]

radiographic and autoradiographic techniques has been of great value for studies on the renewal rate of bone salt in different osteons of compact bone. The information has been expanded by histological staining to disclose the location of the endosteum and cells, by microradiography to show the density of the mineral deposits, and by ultraviolet microscopy to reveal administered fluorescing substances incorporated in the new lamellae. All of this information can be obtained from a single section (Fig. 10.1).

When $Ca^{45}$ or $P^{32}$ is administered intravenously to an animal, it is demonstrable by autoradiography within a few minutes at specific locations. These locations are chiefly the new and incompletely calcified osteons incident to the remodeling of compact bone although rapid deposition also occurs subperiosteally and subendosteally in the growing bones of young animals. Dep-

TABLE 10.1 (*Continued*)

| | Interval after injection of $Sr^{90}$ (10 μC) intraperitoneally | | | | |
|---|---|---|---|---|---|
| Organs | 1, 2, 4, 8 days | 16 days | 32 days | 64 days | 128 days |
| Viscera | No increased density | | | | |
| Skull | Very small decrease in density since 6 hours | Approx. same as 8 days | Density further decreased | Lower density than 32 days | Density same as 64 days |
| Incisors | Still very high density | Still very high density but localized to upper halves | Dense spot only in apices | Dense spot disappeared | As above |
| Spine | Very small decrease in density since 6 hours | Approx. same as 8 days | Density further decreased | Lower density than 32 days | As above |
| Tail | As above | As above | As above | As above | As above |
| Ribs, Pubic bone, Tracheal rings | As above | As above | As above | As above | As above |

osition occurs by ion transfer, a term applied to the movement of ions without specifying the mechanism. When deposition occurs as a one-for-one exchange without a net change in the solid phase, it is called ion exchange, which may be isoionic or heteroionic. There is no essential difference between ion transfer and ion exchange in bone; both occur passively in an approach to equilibrium with the fluids with which the crystal surfaces are in contact.

The most striking sequel of bone-seeking radioisotopes is their deposition in new and incompletely mineralized osteons. The result is a spotty distribution of discrete foci of intense concentrations; these are the areas frequently given the designation of hotspots. Approximately one-half of deposited radiocalcium is diffusely distributed throughout the mineral of bone and produces a rather uniform darkening of autoradiograms, at a much lower

intensity, in contrast with the hotspots. This diffuse component is responsible for long-term exchange and remains buried in the mineral for long periods of time.

## 2.2. Quantitative Autoradiography

In this method the bone section is either placed in contact with stripping film or dipped in emulsion. On subsequent development of the emulsion, all regions corresponding to localization of the radioactive element are darkened. The histological section and the developed emulsion are examined microscopically as a unit; quantitative determination can then be made by densitometry, track counting, or grain counting.

### *2.2.1. Densitometry*

One of the earliest reports on the use of autoradiographs in the quantitative determination of radiation dosages in bone was that by Dudley and Dobyns,[16] whose method consisted of a microphotometric comparison of the blackening by a radioactive sample with a standard source. With high-resolution emulsion, its application is limited owing to the high optical densities required. However, with contrast film the emulsion densities become high enough to allow measurements. The bone sections are usually thicker than the range of the ionizing particles so that the measured densities bear a direct relationship to the total radioactivity at the bone surface.[17] An example of good correlation of results by assay with that of autoradiographic density is shown in Fig. 10.2. The values represent the optical density and the counts per minute of $Ca^{45}$ as measured with a Geiger counter on a sample taken from the region used for the densitometric measurement.

A different approach in densitometry is the use of a matrix theory, i.e., calculation of the geometric relationships between histological section and emulsion with cross-over radiation. The matrix formation increases the understanding of exposure mechanisms, practical performance of quantitative autoradiography, selection of autoradiographic technique, and problems of autoradiographic resolution.[18]

### *2.2.2. Track and Grain Counting*

At high magnifications the tracks or grains are counted for estimation of photographic density. The number of grains is proportional to the amount of radioactivity in the sample: a count of 10 grains/100 $\mu^2$ above background is statistically significant.[11] Alpha particles produce well-defined tracks in nuclear emulsions; each track may serve to identify a single disintegrating atom so that it is possible to calculate the dose rate from the number of tracks per unit area in the emulsion. However, track counting for beta particles

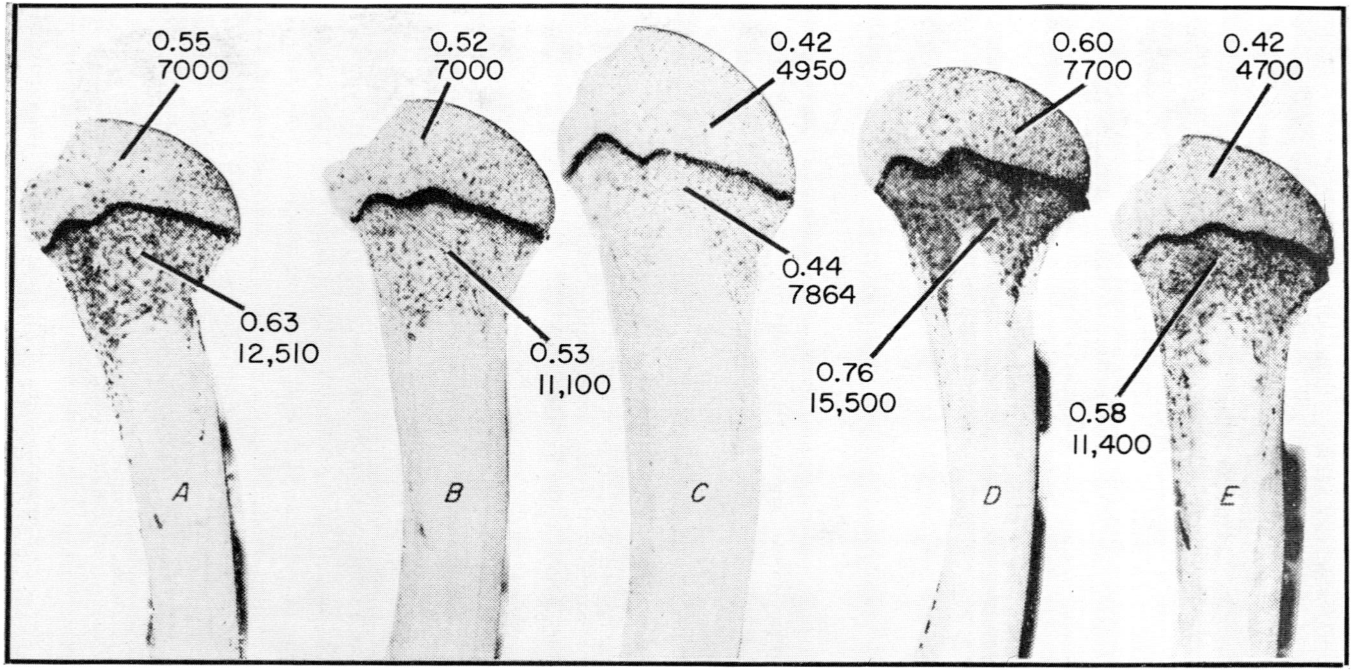


FIG. 10.2. Relation of optical densities (values less than one) and concentrations of $Ca^{45}$ (expressed as counts per minute) as determined by radioassay in comparable areas of sheep femurs. (From Comar, *Radioisotopes in Biology and Agriculture*.[11])

is more difficult: measurement is done either in terms of grain count or by comparison of the photographic blackening with a calibrated source. The method is tedious, and many attempts to introduce automation have been made with varying degrees of success.[12]

Apart from absolute quantitation of radiation dosages to bone, the track- or grain-counting method has been successfully used in the estimation of fine differences in intensity between cells or parts of cells. Application of this method is especially useful for the tritiated thymidine used in studies of chondrogenesis and osteogenesis.

## 3. Radioassay by Counting Methods

The general procedures for assay of bone tissue and other biological samples are given in detail by Comar,[11] Kamen,[19] and Francis *et al.*[20] The methods suitable for a specific analysis will depend upon the element involved, the nature of the sample, and the degree of accuracy required. Certain fluid samples (e.g., blood, plasma, urine, bile, or milk) may under certain conditions be assayed directly by liquid counting. Other samples may require ashing (by either a dry or wet method) to render them homogeneous or to reduce them in volume, or they may require evaporation (in the case of a fluid) to the dry state.

There are many varieties of vessels or holders for material to be counted. In every case the geometry of the system to be used and the self-absorption of the radioactivity by the sample must be taken into account. Moreover, as will be seen, expression of the data may require chemical analysis of the total amount of the element under consideration in order to relate radioactivity to the stable as well as to the radioactive isotopes present.

When a radioactive sample is placed in proximity to a counting system (see Chapter 4), it will make itself known by the response of the measuring device to an ionizing event. This is usually registered as counts per unit of time, e.g., counts per minute (cpm). This quantity will be directly proportional to the amount of radioisotope in the sample when corrections required by background radiation and by the system employed have been made. Background radiation may be determined directly by taking readings from the system in the absence of the sample. Other corrections, e.g., for self-absorption, scattering, placement of counter and sample (geometry), must also be taken into account, by the use of standardized procedures, and, when necessary, by comparison with standards of known radioactivity observed under identical conditions.

Except for purposes of comparison, radioactivity expressed only in terms of counts per minute when counts are made under arbitrary conditions is meaningless. To give meaning to counting data, it is necessary to record them in

relation to a standard of reference. This introduces the concept of *specific activity*, which relates the counting data to the total amount of the element (stable plus radioactive) in the sample. When used with precision, this method designates the amount of the radioactive isotope of the element per unit weight of the element present.

Another method in common use is to relate the amount of the radioactive element present in the sample to the total dose administered. Thus, if the femur of a rat is assayed for radioactivity, the resulting data may be expressed as the per cent of the total dose found at the time of sampling in the whole femur. The procedure to be used will vary according to the needs of the situation. Whenever a short-lived isotope is under observation, correction must be made for the decay estimated to have taken place before the sample is counted.

## 4. Radioisotope Scanning of Bone

Radioisotope scanning at the surfaces of the body has as its principal usefulness the detection, localization, and outlining of lesions in the internal organs. For these purposes there is required the administration of a suitable radioisotope, which will deposit selectively and will result in the emission of gamma rays which can be recorded by a collimated scintillation detector, usually moved from point to point over the area under examination. As an example, one of the early applications of radioisotope scanning was for localization of thyroid tissue with the aid of $I^{131}$.

Scanning of bone at the body surfaces requires the use of emission of penetrating radiation which may be accomplished by one of three methods: (1) the employment of radioisotopes which themselves emit primary gamma radiation; (2) the production of gamma radiation as a result of positron annihilation radiation; or (3) the use of *bremsstrahlung*, comparable to X-rays, produced by the influence of $\beta^-$ particles entering the electrical field around the nucleus of an atom.

The most frequently used procedure for scanning over soft tissues is that of recording gamma radiation produced by positron annihilation. This method depends upon administration of a radioisotope which emits positively charged beta particles ($\beta^+$; positrons). These travel only a short distance in tissues and are absorbed almost immediately. When the positron has spent its kinetic energy in the absorbing medium, it pairs with an electron; both are annihilated, and their energy of mass is converted to a pair of oppositely directed gamma rays of approximately 0.5 Mev energy each (see Fig. 4.2). These rays may be detected and recorded by means of scintillation counters, in the same manner as if they were primary gamma rays.

A less frequently used method employs bremsstrahlung. These rays are

electromagnetic in nature, as are gamma rays, but they resemble X-rays more closely. When fast-moving electrons enter the electrical field around a nucleus, the electron is decelerated; in this process some of the energy is radiated as electromagnetic radiation called bremsstrahlung. Although limited in clinical usefulness because only small amounts of energy reach the surface of the body, the energy emanating from $P^{32}$ is measurable over the skeleton and is most marked in bone lesions, such as those of fractures or of Paget's disease.[21] The history and usefulness of this method have been reviewed by Mehl.[22]

Mulry and Dudley[23] pioneered in the localization of radioactive gallium ($Ga^{72}$) in bone tumors. They placed a small-bore end-window Geiger tube, jacketed by a lead shield, over the suspected area, and obtained a counting rate from the gamma rays emitted by the isotope. Subsequently, Dudley *et al.*[24] improved on the method by the use of better collimation and scintillation crystal counters. While the use of gallium has been superseded by other gamma emitters better suited to these purposes, this work was an important step in the development of skeletal scanning.

Bauer and Ray[25] reported on the kinetics of strontium metabolism in man, having utilized a collimated scintillation crystal and having combined surface counting with assay of the radioactivity of blood, urine, and feces. Their results were submitted to analysis with an electronic analog computer. Bauer and Wendeberg[26] described the use of external counting of $Ca^{47}$ and $Sr^{85}$ in various localized bone lesions in human subjects.

Corey *et al.*[27] have reported on the application of a high-energy-gamma scanner (HEG scanner) to the detection of bone metastases in patients with breast cancer, where the usual roentgenograms have not revealed the bone lesions (see Chapter 14).

The use of external counting of $Ca^{47}$ and $Sr^{85}$ in fractures has been described; the uptake of these tracers in human tibial and femoral fractures was several times that in normal bone tissue.[26] Similar findings are reported in vertebral fractures in man.[28] To these reports, Wendeberg[29] has added a detailed description of the observations made on patients with fractures of one tibia by using scintillation detectors over the thighs, the knees, and the tibias. The uninjured leg served as the control in each instance (see Chapter 14).

Successful application of the positron annihilation technique to the scanning of bone—by the use of $F^{18}$—has been reported by Blau *et al.*[30] This isotope emits a 0.65 Mev positron and has a half-life of 109.7 minutes. The positron annihilation radiation provides gamma radiation suitable for scanning. The short half-life has the effect of shortening the period over which the subject need be irradiated.

## 5. Whole-Body Counting

The history of measurement of the natural radioactivity of the human body, coupled with detection and recording of the smallest possible increase that may occur from occupational or other causes, has been reviewed by Spiers[31]

TABLE 10.2
DEVELOPMENT OF WHOLE-BODY COUNTING SINCE 1929[a]

| Date | Experimenter and apparatus | Subject observation time | Approximate limit of detection (or standard error): As μg Ra equivalent[b] | Approximate limit of detection (or standard error): As per cent of total body potassium |
|---|---|---|---|---|
| 1929 | Schlundt *et al.*: Ion chamber at 1 atm | — | ~ 5 | — |
| 1933 | Schlundt *et al.*: Ion chamber at 1 atm | — | ~ 0.2 | — |
| 1937 | Evans: Geiger-Müller tube | — | 0.1 | — |
| 1947 | Hess and McNiff: Ion chamber at 1 atm | — | 0.03 | — |
| 1951 | Sievert: High-pressure ion chambers | 2 hours | 0.005 (SE) | ~ 50 (SE) |
| 1953 | Burch and Spiers: Differential high-pressure ion chambers | 2 hours | 0.003 (SE) | ~ 30 (SE) |
| 1956 | Sievert: High-pressure ion chambers underground | 3–4 hours | 0.001 (SE) | ~ 10 (SE) |
| 1957 | Los Alamos: 4 π liquid scintillator | 15 minutes | ~ 0.0001 (SE) | 1 (SE) |
| to | Argonne National Laboratory: NaI scintillator | 15 minutes | ~ $0.0003_5$ (SE) | 3.5 (SE) |
| 1960 | Leeds: Plastic scintillator apparatus | 15 minutes | ~ $0.0001_5$ (SE) | 1.5 (SE) |

[a] From Spiers, *Whole-Body Counting.*[31]
[b] By "μg Ra equivalent" is meant micrograms of Ra with all daughter products at equilibrium.

(Table 10.2). Beginning with the utilization of ionization chambers in 1929 and followed by employment of Geiger-Müller counters in 1937, the methodology has since been greatly improved, both as to sensitivity and accuracy, by

the employment of large-scale scintillation apparatus. The assembly in common use, pioneered by Miller and Marinelli,[32] utilizes sodium iodide crystals and has largely superseded both the ion-chamber and Geiger-counter instruments. An account of the state of whole-body counting is available in the *Proceedings of the Symposium on Whole-Body Counting*, held by the International Atomic Energy Agency in Vienna in 1961.[33]

Maximum performance of a whole-body counting facility depends upon the extent and design of the crystal array, the effectiveness of the shielding used, and the high-capacity multichannel analyzers. An important consideration is

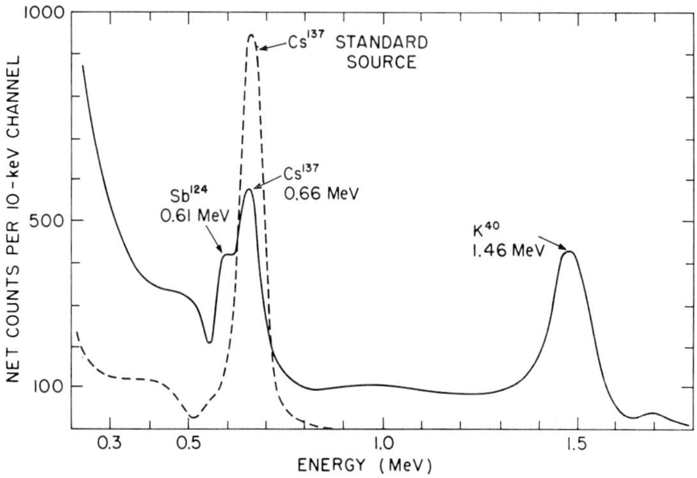


FIG. 10.3. Gamma-ray spectrum from total-body counting of subject. (From MacDonald, *Whole-Body Counting*.[34])

the achievement of a high ratio of the counting rate to background radiation. These characteristics provide for high sensitivity and for the identification of gamma-emitting radioisotopes from the spectrum recorded by the analyzer. A test of a particular facility is its record of the peak produced by the content of naturally radioactive potassium, $K^{40}$, which occurs at 1.46 Mev. Figure 10.3 is a reproduction of the gamma-ray spectrum obtained[34] from a subject who shows, in addition to the normal $K^{40}$ peak and the usual peak for $Cs^{137}$ at 0.66 Mev attributable to fallout, an unexpected distortion at 0.61 Mev resulting from an accidental inhalation of a minute amount of $Sb^{124}$.

Whole-body counting, of course, does not of itself discriminate between radioisotopes deposited in the skeleton and those in the soft tissues. When, however, virtually all of a retained isotope is known to be held in the bones, as is the case with radium, the total body measurement may be taken as

equivalent to the skeletal content. Also, as noted in Chapter 7, direct determination of isotope retention by means of whole-body counting obviates the necessity of deriving a figure for retention by subtracting the amount of isotope excreted in the urine and feces from that of the intake and thus simplifies the recording of retention at desired intervals. Figure 10.4 illustrates the retention of $Ca^{47}$ in the body, recorded from whole-body counts and compared with data obtained by collection and counting of excreta.[35]

It is noted by Spiers[31] that the attainment of maximum performance, as defined above, is not necessarily the answer to every problem and that there

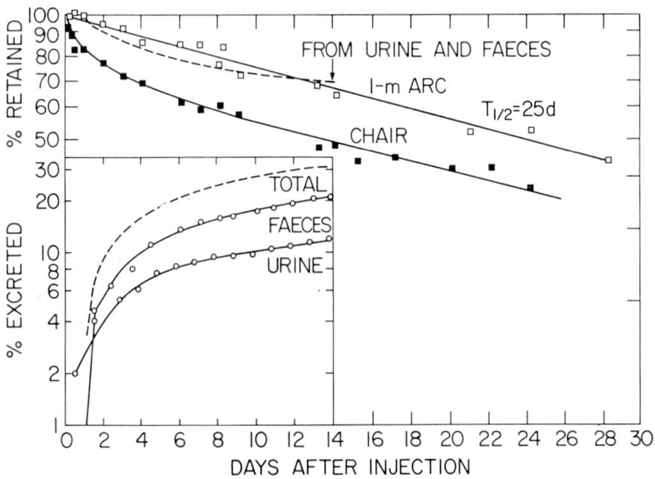


FIG. 10.4. Whole-body counts of $Ca^{47}$ retention in chair and 1-m arc from a counting device, with excreta collected and counted for comparison. (From Sargent, *Whole-Body Counting*.[35])

is also interest in what can be achieved at various levels of elaboration of apparatus. Veall[36] has described methods for clinical counting with simple systems; such methods are best used in combination with metabolic balance studies. Veall has used a method which requires only that eight Geiger-Müller tubes be tied to a frame—the subject either standing in the middle of the frame or in a horizontal position in bed. Since in such a system the background radiation is unshielded, the method requires a proportionately larger dose of the radioisotope than is needed in a system which gives maximum performance. For present purposes, it need only be noted that the level of performance is correlated with cost, and only in certain selected instances is it necessary to make large investments for apparatus more elaborate than that needed for routine external counting procedures.

Another factor of importance in whole-body counting is the geometry of the

system. The two arrangements in most common use are: (1) the subject sits in the chamber in a chair, or (2) he lies in the arc of a circle with a radius of 1 m as measured from the counting device.[35] These two positions give somewhat different results, partly owing to the effect of the geometry on the counting system and partly because of such physiological factors as the time required for mixing of the administered radioisotope in the exchangeable pool.

An extensive review of low-level gamma-ray spectrometry, with special attention to whole-body counting, has been published by Marinelli *et al.*[37] They have described both the methods and results of application of ultrasensitive gamma-ray detectors and predict that the forthcoming improvements in sensitivity will be rewarding to many fields of investigation.

## References

1. H. L. Duhamel, Sur le Développement et la Crûe des Os des Animaux, *Hist. Acad. Roy. Sci.*, **55**: 354-370 (1742).
2. J. Belchier, An Account of the Bones of Animals Being Changed to a Red Colour by Aliment Only, *Phil. Trans. Roy. Soc. London*, **39**: 287-288 (1736).
3. J. Belchier, A Further Account of the Bones of Animals Being Made Red by Aliment Only, *Phil. Trans. Roy. Soc. London*, **39**: 299-300 (1736).
4. J. C. Aub, L. T. Fairhall, A. S. Minot, and P. Reznikoff, *Lead Poisoning*, Williams & Wilkins Co., Baltimore, 1926.
5. J. Vincent, Microscopic Aspects of Mineral Metabolism in Bone Tissue with Special Reference to Calcium, Lead and Zinc, *Clin. Orthopaed.*, **26**: 161-175 (1963).
6. R. A. Milch, D. P. Rall, and J. E. Tobie, Fluorescence of Tetracycline Antibiotics in Bone, *J. Bone Joint Surg.*, **40-A**: 897-910 (1958).
7. O. Chiewitz and G. Hevesy, Radioactive Indicators in the Study of Phosphorus Metabolism in Rats, *Nature (London)*, **136**: 754-755 (1935).
8. D. M. Greenberg, Studies in Mineral Metabolism with the Aid of Artificial Radioactive Isotopes. VIII. Tracer Experiments with Radioactive Calcium and Strontium on the Mechanism of Vitamin D Action in Rachitic Rats, *J. Biol. Chem.*, **157**: 99-104 (1945).
9. G. A. Boyd, *Autoradiography in Biology and Medicine*, Academic Press Inc., New York, 1955.
10. H. Yagoda, *Radioactive Measurements with Nuclear Emulsions*, John Wiley & Sons, Inc., New York, 1949.
11. C. L. Comar, *Radioisotopes in Biology and Agriculture*, McGraw-Hill Book Co., Inc., New York, 1955.
12. Conference on Autoradiography, *Lab. Invest.*, **8**: 57-333 (1959).
13. A. Engström, R. Björnerstedt, C-J. Clemedson, and A. Nelson, *Bone and Radiostrontium*, Almquist & Wiksell, Stockholm, 1957.
14. S. Ullberg, Autoradiographic Studies on the Distribution of Labelled Drugs in the Body, in *Proc. Second Intern. Conf. Peaceful Uses of Atomic Energy*, Vol. 24, pp. 248-254, United Nations, Geneva, 1958.
15. R. Amprino, Rapporti fra Processi di Ricostruzione e Distribuzione dei Minerali Nelle Ossa. II. Ricerche con Metodo Autoradiografico, *Z. Zellforsch. Mikroskop. Anat.* **37**: 240-273 (1952).

16. R. A. Dudley and B. M. Dobyns, The Use of Autoradiographs in the Quantitative Determination of Radiation Dosages from $Ca^{45}$ in Bone, *Science*, **109**: 327-328 and 342 (1949).
17. W. P. Norris and L. A. Woodruff, The Fundamentals of Radioautography, *Ann. Rev. Nucl. Sci.*, **5**: 297-326 (1955).
18. E. Odeblad, Contributions to the Theory and Technique of Quantitative Autoradiography with $P^{32}$ with Special Reference to the Granulosa Tissue of the Graafian Follicles in the Rabbit, *Acta Radiol., Suppl. 93*, 123 pp. (1952).
19. M. D. Kamen, *Isotopic Tracers in Biology*, 3rd ed., Academic Press Inc., New York, 1957.
20. G. E. Francis, W. Mulligan, and A. Wormall, *Isotopic Tracers*, 2nd ed., The Athlone Press, University of London, London, 1959.
21. M. Tubiana, P. Albarede, and H. Nahum, Study of Radioactive Phosphorus ($P^{32}$) Distribution in Man by External Bremsstrahlung Measurements, in *Proc. Second Intern. Conf. Peaceful Uses of Atomic Energy*, Vol. 26, pp. 217-223, United Nations, Geneva, 1958.
22. H. G. Mehl, The Distribution of a Pure Beta-emitter in the Human Body. Problems and Preliminary Results of Bremsstrahlung Measurements in vivo, in *Medical Radioisotope Scanning*, pp. 125-142, Intern. Atomic Energy Agency, Vienna, 1959.
23. W. G. Mulry and H. C. Dudley, Studies of Radiogallium in Bone Tumors, *J. Lab. Clin. Med.*, **37**: 239-252 (1951).
24. H. C. Dudley, H. A. Markowitz, and T. G. Mitchell, Studies of the Localization of Radioactive Gallium ($Ga^{72}$) in Bone Lesions, *J. Bone Joint Surg.*, **38-A**: 627-637 (1956).
25. G. C. H. Bauer and R. D. Ray, Kinetics of Strontium Metabolism in Man, *J. Bone Joint Surg.*, **40-A**: 171-186 (1958).
26. G. C. H. Bauer and B. Wendeberg, External Counting of $Ca^{47}$ and $Sr^{85}$ in Studies of Localized Skeletal Lesions in Man, *J. Bone Joint Surg.*, **41-B**: 558-580 (1959).
27. K. R. Corey, P. Kenny, E. Greenberg, and J. S. Laughlin, Detection of Bone Metastases in Scanning Studies with Calcium-47 and Strontium-85, *J. Nucl. Med.*, **3**: 454-471 (1962).
28. G. C. H. Bauer and P. Scoccianti, Uptake of $Sr^{85}$ in Non-malignant Vertebral Lesions in Man, *Acta Orthopaed. Scand.*, **31**: 90-102 (1961).
29. B. Wendeberg, Mineral Metabolism of Fractures of the Tibia in Man Studied with External Counting of $Sr^{85}$, *Acta Orthopaed. Scand., Suppl. 52*, 79 pp. (1961).
30. M. Blau, W. Nagler, and M. A. Bender, Fluorine-18: A New Isotope for Bone Scanning, *J. Nucl. Med.*, **3**: 332-334 (1962).
31. F. W. Spiers, Whole-body Counting: an Introductory Review, in *Whole-body Counting*, pp. 3-12, Intern. Atomic Energy Agency, Vienna, 1962.
32. C. E. Miller and L. D. Marinelli, Retention of Radium in Humans from 20 to 29 years after Intravenous Administration, in *Whole-Body Counting*, pp. 279-284, Intern. Atomic Energy Agency, Vienna, 1962.
33. *Whole-body Counting*, Intern. Atomic Energy Agency, Vienna, 1962.
34. N. S. MacDonald, Recent Uses of a Total-body Counter Facility in Metabolic Research and Clinical Diagnosis with Radionuclides, in *Whole-body Counting*, pp. 502-514, Intern. Atomic Energy Agency, Vienna, 1962.
35. T. W. Sargent, Metabolic Studies with $Fe^{59}$, $Ca^{47}$ and $C^{11}$ in Various Diseases, in *Whole-body Counting*, pp. 447-468, Intern. Atomic Energy Agency, Vienna, 1962.

36. N. Veall, Clinical Whole-body Counting with Simple Detection Systems, in *Whole-body Counting*, pp. 397-405, Intern. Atomic Energy Agency, Vienna, 1962.
37. L. D. Marinelli, C. E. Miller, H. A. May, and J. E. Rose, Low Level Gamma-ray Scintillation Spectrometry: Experimental Requirements and Biomedical Applications, *Advan. Biol. Med. Phys.*, **8**: 81-160 (1962).

# Chapter 11

# *Radioisotopes and the Organic Portion of Bone*

## 1. Introductory Statement

The ions participating in mineralization of bone, notably $Ca^{2+}$, $HPO_4^{2-}/PO_4^{3-}$, and $CO_3^{2-}$, have a special affinity for the collagen fibers of bone, an affinity which expresses itself in the phenomenon of calcification, beginning with the nucleation of crystals of a mineral of the apatite series and continuing with crystal growth. The physiological analogs of $Ca^{2+}$, and especially the heavier of the alkaline earths, including $Sr^{2+}$, $Ba^{2+}$, and $Ra^{2+}$, have properties in common with $Ca^{2+}$ in that they are all divalent and may substitute for this ion in the mineral of bone. The use made of the isotopes of the elements native to bone and of their analogs in the study of calcification has been explored in Chapters 5 and 7.

Other elements that may be grouped together because of an affinity for bone have quite different properties. With the exception that they are regarded as bone seekers, these elements have little in common with the alkaline earths, and they do not associate themselves with the mineral of bone in a specific manner. Their patterns of deposition and distribution are unlike those of the alkaline earths and, to a large extent, of each other, except that there are subgroups the members of which exhibit similarities. These elements associate themselves with the organic matrix and the surfaces of bone without contributing to the mineralization of collagen fibers. Their special characteristics, and such use as may be made of them in the study of bone, form the subject matter of this chapter.

## 2. The Matrix and Surfaces of Bone

### 2.1. Plutonium

Plutonium heads the list of elements foreign to the organism but with an affinity for bone; it is a man-made element which is used in nuclear weapons and has a potential industrial hazard. Of the 16 isotopes of this element that have been characterized, interest centers in the fissionable $Pu^{239}$ which possesses a half-life of 24,400 years.[1] The biological behavior of $Pu^{239}$ has been intensively studied at the Radiobiology Laboratory at the University of Utah, and the descriptions to follow are drawn largely from the work of this group.[2]

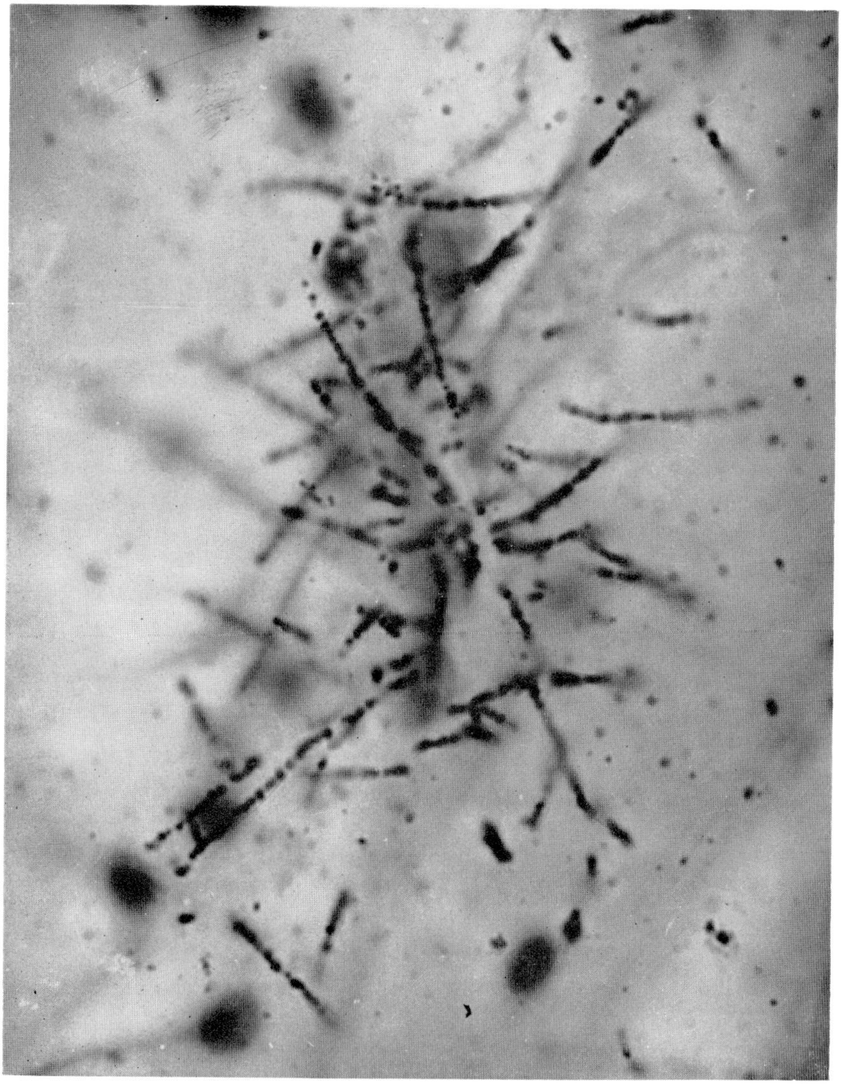

FIG. 11.1. Detailed autoradiograph of an osteoclast found to be a source of radioactivity along the surface of a trabecula in the vertebral body of a rat sacrificed at 24 hours following injection of $Pu^{239}$. Illustrates tracks of alpha particles emitted from plutonium concentrated in the osteoclast. (Magnification × 1560.) (From Arnold and Jee, *Am. J. Anat.*[5])

$Pu^{239}$ is unique in that it has four valence states—2+, 3+, 4+, and 6+—all of which can coexist in aqueous solution. Its small atomic size and high charge result in extensive hydrolysis and complex ion formation. Plutonium-239 is virtually insoluble at pH 7, and when injected intravenously in the tetravalent state it is rapidly bound to protein to such an extent that less than 10% is diffusible. Its removal from blood is very much slower than that of radium, owing to the differences in their chemical states.[3]

Jee and Arnold[4] have studied the distribution of $Pu^{239}$ in bone as well as the radiation dosage to the cellular elements, and the radiotoxicity resulting therefrom. The radioisotope was given to young beagles as plutonium (4+) citrate in a single intravenous injection of approximately 2.7 μc of $Pu^{239}$ per kilogram of body weight and observations were made at from 1 to 1576 days thereafter. Plutonium, demonstrable by autoradiography, was found to be deposited chiefly in the periosteal and endosteal surfaces of bone, including the surfaces of metaphyseal trabeculae and haversian canals. When bone formation increased at endosteal surfaces, the initial deposit of plutonium was found buried beneath the new bone. The sites containing the highest concentrations of plutonium and those areas which exhibited a relatively high rate of remodeling were found to be the earliest to show radiation damage. It was concluded that, in the case of alpha-emitting bone-seeking radioisotopes, the dose delivered to the cells lining bone played the dominant role in induction of bone changes. Changes occurred early in trabecular bone and later in cortical bone; changes were manifested by plugging of haversian canals and resulted in the death of osteocytes. A still later effect was the induction of osteogenic sarcomas.

An interesting by-product of the studies of Arnold and Jee[5] on the distribution of $Pu^{239}$ in the skeletons of rats has been their contribution to the knowledge of bone growth and osteoclastic activity. They found that plutonium localizes in high concentrations at endosteal surfaces and to a lesser degree at periosteal and endosteal surfaces of vascular channels. As described above for beagles, the initial deposit of plutonium was buried by bone. Where resorption occurred, the plutonium localized in the resorbed bone was progressively concentrated in osteoclasts and was later transferred to macrophages. The localization of plutonium in bone and in histiocytes, together with its lack of diffusibility, is in keeping with the colloidal nature of its solutions (Fig. 11.1).

### 2.2. Thorium

As noted in Chapter 3, transmutation of $Ra^{228}$ (mesothorium) via $Ac^{228}$ to $Th^{228}$ (radiothorium) results in the formation of an element which loses the properties of radium and assumes those of thorium. $Th^{228}$, when ad-

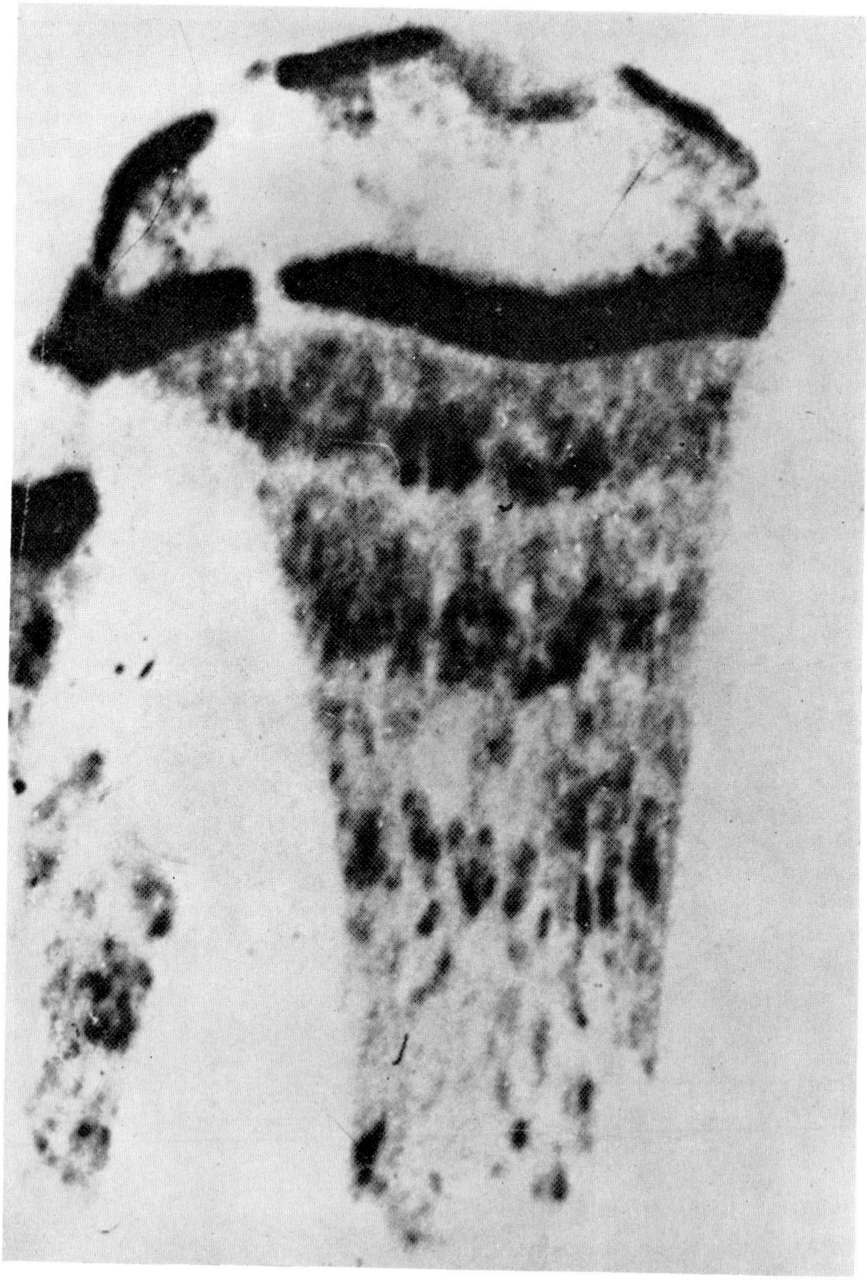

FIG. 11.2. Autoradiograph of proximal end of tibia and fibula from a 49-day-old rat, showing stratification of radiosulfate in the metaphysis as a result of periodic administra-

ministered as such, is retained in the skeleton in much the same manner as $Pu^{239}$, but it decays to $Ra^{224}$, which again has the properties of radium. The influence of these successive decays and transmutations on translocations in the thorium series is considered in Chapter 3. Here it may be repeated that the radium daughters of $Th^{232}$ behave like $Ra^{226}$ in the organism and thus resemble their physiological analog, $Ca^{2+}$, while thorium daughters in the same series behave as elements foreign to the skeleton, as does plutonium.

## 2.3. The Lanthanons

Elements 57 to 71, lanthanum to lutetium inclusive, are also without physiological analogs, but all are fixed in the skeleton once they gain access to the circulation.[6] According to their behavior in the animal organism, they may be divided into the same three groups familiar to lanthanon chemists: (1) the light lanthanons—lanthanum, cerium, praseodymium, neodymium, promethium, and samarium; (2) the transition group—europium, gadolinium, and terbium; and (3) the heavy lanthanons—dysprosium, holmium, erbium, thulium, ytterbium, and lutetium.

Absorption of all the lanthanons from the gastointestinal tract is poor. Cerium ($Ce^{144}$) is of interest because of its relatively large contribution to the fission products. $Ce^{144}$, like plutonium, is deposited on endosteal and periosteal surfaces and in the vicinity of small blood vessels in compact bone. Its deposition in uncalcified osteoid tissue of rachitic rats indicates a specific affinity for bone protein, even in the absence of mineral.

Although all of the lanthanons are similar in their affinity for the bone matrix, the heavy members of the lanthanon series exhibit a greater deposition in the skeleton, and there are suggestions that differences in the behavior of the light and heavy lanthanons may be determined by the type or strength of the chemical combination of the lanthanons with the bone protein. The greater stability of the chelates of the heavy lanthanon may be the basis for their greater skeletal deposition.

## 2.4. Yttrium

The prototype of elements depositing preferentially in the surfaces of bone, especially on irregular surfaces where osteoclastic activity is in progress or has occurred, is yttrium.[7] This element has been studied both as $Sr^{90} + Y^{90}$ in equilibrium[8] and as $Y^{91}$. Ramsden[9] has reviewed the experimental work on

tion. The rat received weekly intraperitoneal injections of 2.0 mg of estradiol-17β benzoate in 0.2 ml of corn oil for 4 consecutive weeks; 24 hours after each dose of estrogen the rat received an intraperitoneal injection of 100 μc of $S^{35}$ as sodium sulfate. The animal was killed 24 hours after the fifth injection of the isotope. Fixation, 10% formalin. (Magnification × 22.) (From Dziewiatkowski *et al.*, *J. Biophys. Biochem. Cytol.*[10])

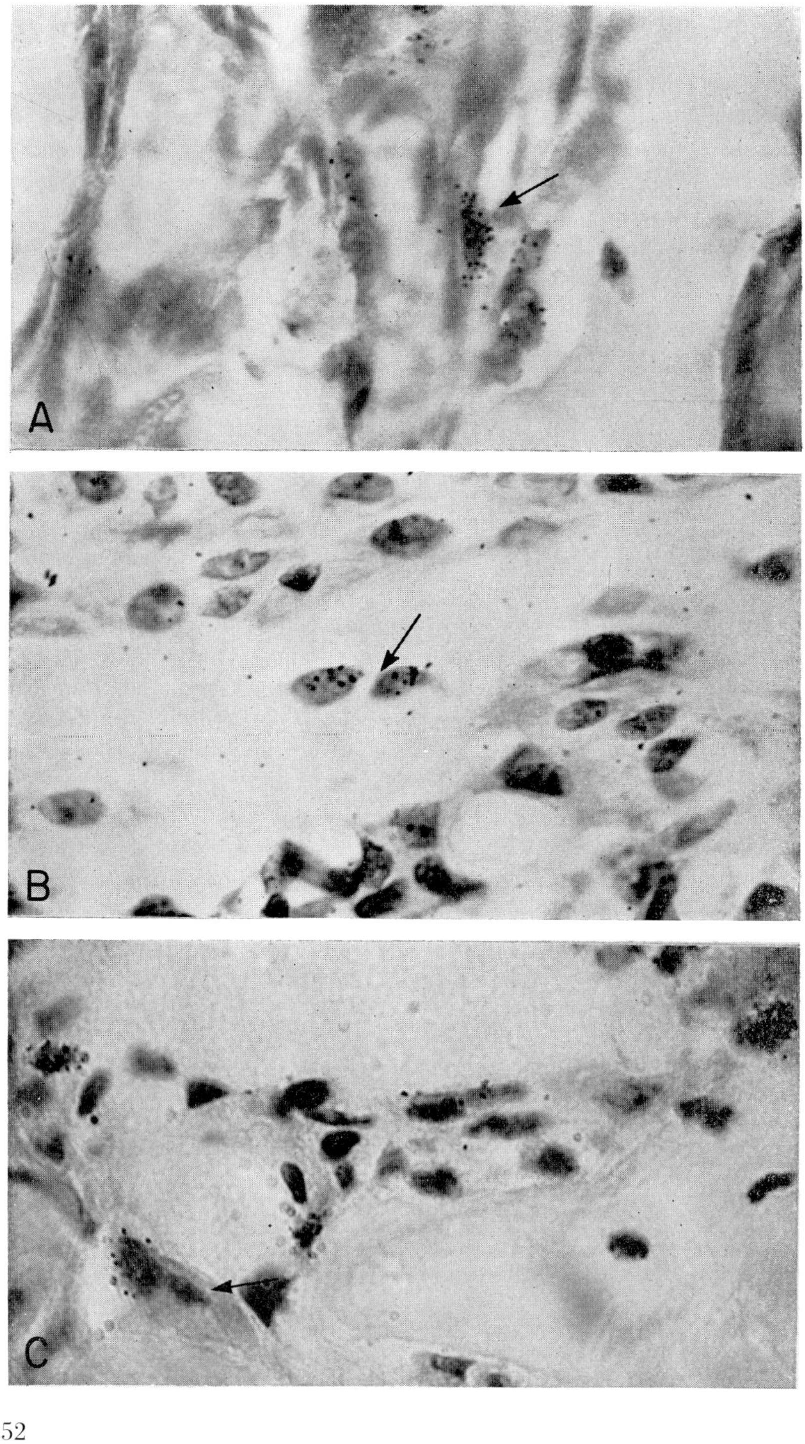
A
B
C

radioyttrium including its distribution in tissues, the mechanism of its deposition in bone, and the state of the element in the blood. Ramsden concludes from her studies on bone that, unlike the alkaline earths, yttrium (and the other rare earths) is taken up on resorbing or quiescent bone surfaces where calcification is not taking place. Yttrium is also present diffusely throughout mineralized bone; the mechanism of deposition is possibly that of surface adsorption.

## 3. Labeled Organic Molecules

Increasing use is being made of labeled organic molecules in the study of the organic portion of bone. This offers a choice of three procedures: (1) a radioactive ion may be introduced into the organism, to be incorporated in a molecule by biosynthesis; (2) a precursor bearing a radioelement may be administered, the whole to be used in the formation of a larger molecule; and (3) an entire molecule, such as a hormone or vitamin, may be labeled before introduction into the organism. In each instance, the interest is in the final product and in obtaining information concerning its formation, distribution, metabolism, physiological effects, and fate.

### 3.1. Biosynthesis

When $S^{35}$ is administered as the sulfate ion ($S^{35}O_4^{2-}$) to young growing animals, the concentration of radiosulfur in epiphyseal cartilage increases over a period of some 24 hours. This is attributed to a rapid utilization of the sulfate-$S^{35}$ by the cartilage for the synthesis of chondroitin sulfate. The subsequent fate of sulfate-$S^{35}$ and its partial utilization in the ground subtance of bone may be traced autoradiographically,[10,11] as shown in Fig. 11.2.

Cartilage is permeable to inorganic sulfate either *in vitro* or *in vivo;* the cells utilize the sulfate almost as rapidly as it enters the tissue.[12] In both systems, the final product isolated from cartilage was chondroitin sulfate labeled in the ester sulfate group.[13] Radiosulfur has also been used as a tracer to observe the repair of fractures in animal experiments.[14]

### 3.2. Precursors

The deposition of matrix material can be studied by chemical assay or by autoradiography following administration of various precursors or by the

FIG. 11.3. Microautoradiograph of sections of spongiosa of femurs of mice following subcutaneous injections of thymidine-$H^3$. The sections, 6 μ thick, were decalcified in EDTA, dipped in Kodak NTB-3 emulsion for 5 minutes and exposed for 10 days, then stained by hematoxylin and eosin. A, Osteoblast; B, osteocytes; C, osteoclast. (Magnification × 1040.) (Courtesy of D. J. Simmons.)

addition of labeled precursors to incubating media in the presence of bone tissue slices. The precursors most frequently used are $S^{35}$-labeled methionine and cystine, tritiated amino acids (i.e., thymidine, glycine, methionine, and leucine), and $C^{14}$-labeled thymidine, glycine, tricarboxylic acid cycle intermediates, and bicarbonate.

### 3.2.1. $S^{35}$-*Labeled Precursors*

Protein synthesis, in areas of intense cell growth, can be visualized autoradiographically with the use of $S^{35}$-labeled methionine and cystine.[15] In very young rats, the preosteoblasts, prechondroblasts, and preodontoblasts are intensely labeled, indicating that these cells are active in protein synthesis. When the cells mature to the secretory phase, the intensity of radioactivity diminishes with time as growth displacement occurs. The turnover rate of labeled material is slower in keratinized and bone tissue than in glandular tissue.

### 3.2.2. *Tritiated Amino Acids*

Tritiated thymidine has come into wide use as a label for cell nuclei. Thymidine enters the structure of deoxynucleic acid (DNA) during interphase. When tritiated thymidine is administered, the nuclei of cells undergoing division are labeled, and they and their daughter cells may be subsequently identified by microautoradiography, as shown in Fig. 11.3. This procedure has been employed in the study of (1) the life cycle and transformations of the cells of bone, including preosteoblasts, osteoblasts, osteocytes, and osteoclasts,[16-21] (2) chondrogenesis,[22] and (3) in cell population kinetics.[20,21]

It has been shown[20] that the incorporation of osteoblasts into bone is not a random process, but it appears that the osteoblast must spend some time in the periosteal surface before becoming either an osteocyte or a relatively inactive osteoblast lining a haversian canal. During its most active period on the periosteal surface, it was estimated that, on an average, an osteoblast produces two or three times its own volume of matrix.

The uptake of thymidine is not always associated with cell division since a considerable fraction of labeled preosteoblasts and a small percentage of

---

FIG. 11.4. Autoradiograph of a living tibia of a rat, implanted in rectus abdominis muscle. One week after implantation, rat was injected with 5.0 μc of $C^{14}$-labeled proline, and sacrificed 1 day later. Radioactivity is concentrated in zone of proliferating cartilage cells, zone of invasion of cartilage by growing bone, and subendosteal and subperiosteal bone. These correspond to the locations in which new bone matrix is being formed. The zone of vesicular cartilage cells which are forming no matrix shows no uptake. (Magnification × 17.5.) (From Ray *et al.*, *Radioisotopes and Bone*.[28])

labeled osteoblasts did not divide. Some labeled cells also become osteocytes without division. This uptake of thymidine may be renewal of DNA which is associated with cellular function and not with mitosis.[21]

Tritiated glycine has proved to be a useful tracer for newly formed bone matrix.[23] In sites of bone growth in adult mice, the glycine label has been found to give a strong reaction in the cytoplasm of osteoblasts within 30 minutes after injection of the precursor, while the preosseous tissue reacts only moderately. At 4 hours the radioactivity in the osteoblasts has diminished, while the preosseous tissue is heavily labeled. From these observations, it has been concluded that osteoblasts elaborate a matrix precursor within their cytoplasm and release it to preosseous tissue.[24]

### 3.2.3. *$C^{14}$-Labeled Precursors*

Direct evidence for a collagen precursor, tropocollagen, has been obtained by the addition of proline-$C^{14}$ to a suspension of osteoblasts. A protein-bound hydroxyproline-$C^{14}$ appeared almost immediately, although collagen fibers were not detectable until a day later.[25,26] Studies with the aid of labeled glycine have suggested that bone collagen, like tendon collagen, has a relatively low rate of turnover.[27] Ray *et al.*[28] have studied the uptake of proline-$C^{14}$, as well as of $Ca^{45}$, by dead and living bone implanted in the rectus abdominis muscle of rats (Fig. 11.4). They were unable to account for the uptake of proline-$C^{14}$ in dead bone—they had used the isotope as an indicator of the biosynthesis of collagen.

Turnover rates of chondroitin sulfate are similar when $C^{14}$-labeled precursors or sulfate-$S^{35}$ are used. Campo and Dziewiakowski[29] incubated bovine costal cartilage in solutions containing sulfate-$S^{35}$, leucine-$C^{14}$, or phenylalanine-$C^{14}$ and reported that the radioisotopes were seen predominantly in the chondrocytes in autoradiograms. The protein-polysaccharides isolated from the slices were radioactive; radiosulfur and radiocarbon were incorporated at roughly parallel rates. They showed that protein as well as the polysaccharide moieties of the protein-polysaccharides are synthesized simultaneously and intracellularly.

The metabolism of amino acids in bone cells is also amenable to study in an *in vitro* system. Vaes and Nichols[30] have shown that glycine-$C^{14}$ in the incubation medium is incorporated intact into the matrix of surviving metaphyseal bone fragments.

## 3.3. Labeled Molecules

Evidence that endosteum is a target for estrogens, characteristic for the mouse, is based on the localization of radioactivity in the skeleton, following administration of the hormone, long after the soft tissues have been cleared of

this activity. The patterns of distribution and metabolism of labeled estrone-16-$C^{14}$ are different from those of estradiol-6,7-$H^3$.[31]

When ergocalciferol-$C^{14}$ (vitamin D) was administered to rachitic rats, tissues intimately connected with the turnover of phosphate such as bone, intestines, kidneys, and striated and cardiac muscle were found to contain significant amounts of the labeled vitamin.[32]

## References

1. D. Strominger, J. M. Hollander, and G. T. Seaborg, Table of Isotopes, *Rev. Mod. Phys.*, **30**: 585-904 (1958).
2. USAEC Progress Reports, University of Utah.
3. B. J. Stover, Metabolism of Radioactive Isotopes Which Deposit Mainly in the Skeleton, *Health Phys.*, **1**: 373-378 (1959).
4. W. S. S. Jee and J. S. Arnold, The Toxicity of Plutonium Deposited in Skeletal Tissues of Beagles. 1. The Relation of the Distribution of Plutonium to the Sequence of Histopathologic Bone Changes, *Lab. Invest.*, **10**: 797-825 (1961).
5. J. S. Arnold and W. S. S. Jee, Bone Growth and Osteoclastic Activity as Indicated by Radioautographic Distribution of Plutonium, *Am. J. Anat.*, **101**: 367-418 (1957).
6. P. W. Durbin, M. H. Williams, M. Gee, R. H. Newman, and J. G. Hamilton, Metabolism of the Lanthanons in the Rat, *Proc. Soc. Exptl. Biol. Med.*, **91**: 78-85 (1956).
7. J. Jowsey, H. A. Sissons, and J. Vaughan, The Site of Deposition of $Y^{91}$ in the Bones of Rabbits and Dogs, *J. Nucl. Energy*, **2**: 168-176 (1956).
8. E. Lloyd, The Relative Distributions of Radioactive Yttrium and Strontium and the Secondary Deposition of $^{90}Y$ built up from $^{90}Sr$, *Intern. J. Radiation Biol.*, **3**: 475-492 (1961).
9. E. N. Ramsden, A Review of Experimental Work on Radio-yttrium Comprising 1. The Tissue Distribution, 2. The Mechanism of Deposition in Bone, and 3. The State in the Blood, *Intern. J. Radiation Biol.*, **3**: 399-410 (1961).
10. D. D. Dziewiatkowski, F. Bronner, N. Di Ferrante, and R. M. Archibald, Some Aspects of the Metabolism of Sulfate-$S^{35}$ and Calcium-45 in the Metaphyses of Immature Rats. Influence of β-estradiol Benzoate, *J. Biophys. Biochem. Cytol.*, **3**: 151-160 (1957).
11. C. P. Leblond, L. F. Bélanger, and R. C. Greulich, Formation of Bones and Teeth as Visualized by Radioautography, *Ann. N. Y. Acad. Sci.*, **60**: 631-659 (1955).
12. R. D. Campo and D. D. Dziewiatkowski, A Consideration of the Permeability of Cartilage to Inorganic Sulfate, *J. Biophys. Biochem. Cytol.*, **9**: 401-408 (1961).
13. D. D. Dziewiatkowski, Isolation of Chondroitin Sulfate-$S^{35}$ from Articular Cartilage of Rats, *J. Biol. Chem.*, **189**: 187-190 (1951).
14. R. B. Duthie, $^{35}S$-sulphate in Bone Repair of Rats, in *Radioisotopes and Bone*, pp. 293-316, Blackwell Scientific Publ., Oxford, 1962.
15. L. F. Bélanger, Autoradiographic Visualization of the Entry and Transit of $S^{35}$ Methionine and Cystine in the Soft and Hard Tissues of the Growing Rat, *Anat. Record*, **124**: 555-580 (1956).
16. D. J. Simmons, Cellular Changes in the Bones of Mice as Studied with Tritiated Thymidine and the Effects of Estrogen, *Clin. Orthopaed.*, **26**: 176-189 (1962).
17. E. A. Tonna, Osteoclasts and the Aging Skeleton: a Cytological, Cytochemical and Autoradiographic Study, *Anat. Record*, **137**: 251-270 (1960).

18. N. F. Kember, Cell Division in Endochondral Ossification. A Study of Cell Proliferation in Rat Bones by the Method of Tritiated Thymidine Autoradiography, *J. Bone Joint Surg.*, **42-B**: 824-839 (1960).
19. R. W. Young, Regional Differences in Cell Generation Time in Growing Rat Tibiae, *Exptl. Cell Res.*, **26**: 562-567 (1962).
20. M. Owen, *Cell Population Kinetics of an Osteogenic Tissue*, I, USAEC Report BNL-6737, Brookhaven National Laboratory, 1963.
21. M. Owen and S. Macpherson, *Cell Population Kinetics of an Osteogenic Tissue*, II, USAEC Report BNL-6738, Brookhaven National Laboratory, 1963.
22. W. M. Rigal, The Use of Tritiated Thymidine in Studies of Chondrogenesis, in *Radioisotopes and Bone*, pp. 197-225, Balckwell Scientific Publ., Oxford, 1962.
23. J. Carneiro and C. P. Leblond, Role of Osteoblasts and Odontoblasts in Secreting the Collagen of Bone and Dentin, as Shown by Radioautography in Mice Given Tritium-labelled Glycine, *Exptl. Cell Res.*, **18**: 291-300 (1959).
24. J. Gross, The Behavior of Collagen Units as a Model in Morphogenesis, *J. Biophys. Biochem. Cytol.*, **2**: 261-274 (1956).
25. S. F. Jackson and R. H. Smith, Studies on the Biosynthesis of Collagen. I. The Growth of Fowl Osteoblasts and the Formation of Collagen in Tissue Culture, *J. Biophys. Biochem. Cytol.*, **3**: 897-912 (1957).
26. R. H. Smith and S. F. Jackson, Studies on the Biosynthesis of Collagen. II. The Conversion of $^{14}$C-L-proline to $^{14}$C-hydroxyproline by Fowl Osteoblasts in Tissue Culture, *J. Biophys. Biochem. Cytol.*, **3**: 913-922 (1957).
27. A. Neuberger and H. G. B. Slack, The Metabolism of Collagen from Liver, Bone, Skin and Tendon in the Normal Rat, *Biochem. J.*, **53**: 47-52 (1953).
28. R. D. Ray, J. Stevens, I. Lyon, and R. E. Rowland, Uptake of $^{45}$Calcium and $^{14}$Carbon-labelled Proline by Dead and Living Bone, in *Radioisotopes and Bone*, pp. 69-83, Blackwell Scientific Publ., Oxford, 1962.
29. R. D. Campo and D. D. Dziewiatkowski, Intracellular Synthesis of Protein-polysaccharides by Slices of Bovine Costal Cartilage, *J. Biol. Chem.*, **237**: 2729-2735 (1962).
30. G. M. Vaes and G. Nichols, Jr., Metabolism of Glycine-1-$C^{14}$ by Bone *in vitro:* Effects of Hormones and Other Factors, *Endocrinology*, **70**: 890-901 (1962).
31. A. M. Budy, Radioactive Oestrogens and Bone, in *Radioisotopes and Bone*, pp. 227-240, Blackwell Scientific Publ., Oxford, 1962.
32. E. Kodicek, Turnover and Distribution of Vitamin D and Its Mode of Action, in *Conference on the Transfer of Calcium and Strontium Across Biological Membranes*, Academic Press, New York (in press).

# Chapter 12

# *Effects of Radiation upon Bones*

## 1. Units of Radioactivity and of Radiation

The measure of activity of a radioactive substance is the curie (c). More specifically, it is that quantity of a radionuclide in which the number of disintegrations per second (dps) is $3.700 \times 10^{10}$. A millicurie (mc) corresponds to $3.7 \times 10^{7}$ dps, a microcurie (μc) to $3.7 \times 10^{4}$ dps, and a micro-microcurie (μμc) or picocurie (pc) to 0.037 dps. The curie unit was officially redefined in 1950 and is no longer referred to in terms of radium.

The effect of radiation on biological systems is determined by the energy transferred to the system. The two units in common use for expressing this energy are the roentgen (r) and the rad. The roentgen unit, based on the ionization produced in air, is the exposure dose and is used for X-rays or $\gamma$ rays of less than 3 Mev quantum energy. The amount of radiation received by a tissue is defined as the energy absorbed per unit mass of the tissue. This radiation is called the absorbed dose and is measured in rad, which is the unit of absorbed dose of radiation in matter, equal to 100 ergs per gram. The rad is the unit of measurement for all types of radiation, both internal or external.[1,2,3,4]

Different kinds and energies of ionizing radiations vary in the dose required

TABLE 12.1
RBE VALUES[a]

1. X-rays, electrons, and positrons of any specific ionization
RBE = 1

2. Heavy ionizing particles

| Average specific ionization (ion pairs per micron of water) | RBE[b] | Average linear energy transfer to water (kev per micron) |
|---|---|---|
| 100 or less | 1 | 3.5 or less |
| 100 to 200 | 1 to 2 | 3.5 to 7.0 |
| 200 to 650 | 2 to 5 | 7.0 to 23 |
| 650 to 1500 | 5 to 10 | 23 to 53 |
| 1500 to 5000 | 10 to 20 | 53 to 175 |

[a] From *Report of UN Scientific Comm. on Effects of Atomic Radiation.*[1]

[b] For practical purposes, an RBE of 10 is applicable to fast neutrons and protons up to 10 Mev and an RBE of 20 to heavy recoil nuclei for whole-body irradiation and the most sensitive critical organs.

to produce a given biological effect. The relative biological effectiveness (RBE) of one radiation with respect to another is defined as the inverse ratio of the respective doses necessary to bring about a given effect. For example, if the RBE of alpha rays is 10, then for such a system, an alpha-ray dose of 0.1 rad will produce the same biological effect as 1 rad of the reference radiation (Table 12.1).

The roentgen equivalent man, *rem,* is now defined in terms of the rad—the dose in rad multiplied by the relevant RBE factor. The rem is that quantity of any type of ionizing radiation which, when absorbed in the human body, produces an effect equivalent to the absorption of 1 r of X-radiation having a linear energy transfer to water of 3.5 kev per micron. The linear energy transfer, LET, is the linear rate of loss of locally absorbed energy by an ionizing particle traversing a material medium; it is expressed in kilo electron volts per micron (kev/μ).

## 2. External Radiation

The bones of adults have a high degree of resistance to radiation; the fetal skeleton is more sensitive. In Table 12.2 are listed the skeletal abnormalities induced by irradiation. The bones of children, and particularly the growth

TABLE 12.2

MAJOR SKELETAL ABNORMALITIES INDUCED IN MAMMALS BY FETAL IRRADIATION[a]

| | |
|---|---|
| General stunting | Spina bifida |
| Reduced skull dimensions | Reduced and deformed tail |
| [b] Skull deformities | Overgrown and deformed feet |
| [b] Head ossification defects | [b] Club feet |
| Vaulted cranium | Digital reductions |
| Narrow head | Calcaneo valgus |
| Cranial blisters | [b] Abnormal limbs |
| [b] Cleft palate | [b] Syndactyly |
| Funnel chest | [b] Brachydactyly |
| Congenital dislocation of hips | [b] Odontogenesis imperfecta |

[a] From *Report of UN Scientific Comm. on Effects of Atomic Radiation.*[1]

[b] These abnormalities have been found in humans exposed *in utero* to radiation and are attributed to the action of radiation.

apparatus of the long bones, are also susceptible to injury. Such injury is not a hazard of ordinary diagnostic roentgenography, but overexposure may result from therapeutic radiation. Uninformed use of fluoroscopic equipment may also be responsible for harmful effects.

Following irradiation of bone tissue, there is no demarcation between normal and irradiated parts. Irradiated bone is more easily infected than normal bone,

especially if necrotic spots are present. Healing of fractures is not significantly altered after moderate doses.[1]

In experimental animals,[5] total body irradiation will usually result in death from doses below those necessary to cause damage to bones. Much higher doses restricted to local areas may, however, cause extensive changes in the bones. The parallel columns of cartilage cells in the epiphyseal cartilage plate become disarranged, the cells swell, and the matrix takes on a mottled and fibrillar appearance. Growth ceases, mainly as a result of interference with the ingrowth of blood vessels and with the replacement of the hypertrophic cartilage. Osteoblasts may disappear and assume the characteristics of reticular cells. If the damage to the growth apparatus is slight, recovery will occur within a few weeks; larger doses may result in permanent impairment of growth. The changes to be expected in growing children are similar to those found in young animals.

Damage to the bones of adults is most often seen following very heavy and localized X-ray treatment or similar radiation from other external sources. Spontaneous fractures of the ribs, of the neck of the femur, and of the jaw have been reported under such circumstances. Serious marrow changes are produced at doses much lower than those approaching the threshold for damage to bone substance. Damage may be done to the vascular channels, both of bone and of bone marrow, at somewhat higher doses than those that affect the hemopoietic function and the formed elements of the marrow.

## 3. Internal Radiation

The biological effectiveness of internal radiation, in addition to being dependent upon its quality and intensity, is determined by a combination of the physical half-life of the radioelement and its biological half-life, which is the time required for the biological system to eliminate one-half of the substance that has been introduced into it. The effective half-life of a radioisotope in a biological system is the resultant of a combination of its physical and its biological half-life. For those elements fixed in bone, the effective half-life in this location may be assumed to be relatively long; for the long-lived isotopes, it may far exceed the length of life of the individual.

The concept of RBE has a limited usefulness for internal radiation, since the biological effectiveness of any radiation depends on many factors in addition to the characteristics of the radiation. These factors include dose, dose-rate, biological half-life, distribution and retention in tissues, and differences in susceptibility to radiation damage. Moreover, the RBE of alpha to beta radiation is dose dependent, changing markedly with the level of exposure. With bone-seeking radioisotopes, there is the additional problem of a non-

TABLE 12.3
MAXIMUM PERMISSIBLE MAINTAINED BODY BURDENS (MPBB) WITH BONE AS ORGAN OF REFERENCE[a]

| Radionuclide | | | | MPBB in total |
|---|---|---|---|---|
| Z | | A | Decay[g] | body, $q$ (μc) |
| 4 | Be | 7 | ε γ | $2 \times 10^3$ |
| 6 | C[b] | 14 | β− | 400 |
| 9 | F | 18 | β+ | 20[c] |
| 15 | P | 32 | β− | 6[f] |
| 16 | S | 35 | β− | 800 |
| 20 | Ca | 45 | β− | 30[f] |
| | Ca | 47 | β− γ | 5[f] |
| 21 | Sc | 46 | β− γ | 60 |
| | Sc | 47 | β− γ | 60 |
| | Sc | 48 | β− γ | 30 |
| 23 | V | 48 | β+ ε γ | 60 |
| 26 | Fe | 55 | ε | $7 \times 10^3$ |
| | Fe | 59 | β− γ | 100 |
| 28 | Ni | 59 | ε | $10^3$[f] |
| | Ni | 63 | β− | 200[f] |
| | Ni | 65 | β− γ | 4 |
| 30 | Zn | 65 | β+ ε γ | 700 |
| | Zn | 69m | β− γ e− | 40 |
| | Zn | 69 | β− | 40 |
| 31 | Ga | 72 | β− γ | 10 |
| 38 | Sr | 85m | ε γ | 70 |
| | Sr | 85 | ε γ | 70 |
| | Sr | 89 | β− | 4[f] |
| | Sr | 90 | β− | 2[f] |
| | Sr | 91 | β− γ | 3 |
| | Sr | 92 | β− γ | 2 |
| 39 | Y | 90 | β− | 3 |
| | Y | 91m | β− γ | 5 |
| | Y | 91 | β− γ | 5[f] |
| | Y | 92 | β− γ | 2 |
| | Y | 93 | β− γ e− | 2 |
| 40 | Zr | 93 | β− γ e− | 100[f] |
| | Zr | 95 | β− γ e− | 30 |
| | Zr | 97 | β− γ | 5 |

[a] From Intern. Comm. Radiological Protection, *Health Physics*.[2]
[b] $CO_2$.
[c] Bone and teeth.
[d] 9.2 hour half-life.
[e] 13 year half-life.
[f] Values whenever bone is the critical organ.
[g] Modes of disintegration and types of radiation: α (alpha decay), β− (negatron), β+ (positron), ε (orbital electron capture, EC), e− (internal conversion electron), γ (gamma ray).

TABLE 12.3 (*Continued*)

| Radionuclide | | | | MPBB in total |
|---|---|---|---|---|
| Z | | A | Decay[g] | body, $q$ (μc) |
| 41 | Nb | 93m | $\gamma\ e^-$ | 200 |
| | Nb | 95 | $\beta^-\ \gamma$ | 80 |
| | Nb | 97 | $\beta^-\ \gamma$ | 10 |
| 43 | Tc | 96m | $\varepsilon\ \gamma\ e^-$ | $10^4$ |
| | Tc | 96 | $\varepsilon\ \gamma$ | $2 \times 10^3$ |
| | Tc | 97m | $\varepsilon\ \gamma\ e^-$ | 700 |
| | Tc | 97 | $\varepsilon$ | $6 \times 10^3$ |
| | Tc | 99m | $\beta^-\ \gamma$ | $10^5$ |
| | Tc | 99 | $\beta^-$ | 500 |
| 44 | Ru | 97 | $\varepsilon\ \gamma\ e^-$ | 900 |
| | Ru | 103 | $\beta^-\ \gamma\ e^-$ | 100 |
| | Ru | 105 | $\beta^-\ \gamma\ e^-$ | 40 |
| | Ru | 106 | $\beta^-\ \gamma$ | 10 |
| 45 | Rh | 103m | $\gamma\ e^-$ | $10^3$ |
| | Rh | 105 | $\beta^-\ \gamma$ | 200 |
| 47 | Ag | 105 | $\varepsilon\ \gamma$ | 200 |
| | Ag | 110m | $\beta^-\ \gamma$ | 40 |
| | Ag | 111 | $\beta^-\ \gamma$ | 60 |
| 49 | In | 113m | $\gamma\ e^-$ | 90 |
| | In | 114m | $\beta^-\ \varepsilon\ \gamma\ e^-$ | 10 |
| | In | 115m | $\beta^-\ \gamma\ e^-$ | 90 |
| | In | 115 | $\beta^-$ | 60 |
| 50 | Sn | 113 | $\varepsilon\ \gamma\ e^-$ | 30[f] |
| | Sn | 125 | $\beta^-\ \gamma\ e^-$ | 7 |
| 51 | Sb | 122 | $\beta^-\ \gamma$ | 40 |
| | Sb | 124 | $\beta^-\ \gamma$ | 30 |
| | Sb | 125 | $\beta^-\ \gamma\ e^-$ | 70[f] |
| 52 | Te | 125m | $\gamma\ e^-$ | 100 |
| | Te | 127m | $\beta^-\ \gamma\ e^-$ | 50 |
| | Te | 127 | $\beta^-$ | 100 |
| | Te | 129m | $\beta^-\ \gamma\ e^-$ | 20 |
| | Te | 129 | $\beta^-\ \gamma\ e^-$ | 40 |
| | Te | 131m | $\beta^-\ \gamma\ e^-$ | 50 |
| | Te | 132 | $\beta^-\ \gamma\ e^-$ | 30 |
| 55 | Cs | 131 | $\varepsilon$ | $8 \times 10^3$ |
| | Cs | 134m | $\beta^-\ \gamma\ e^-$ | 600 |
| | Cs | 134 | $\beta^-\ \gamma$ | 200 |
| | Cs | 135 | $\beta^-$ | 400 |
| | Cs | 136 | $\beta^-\ \gamma$ | 400 |
| | Cs | 137 | $\beta^-\ \gamma\ e^-$ | 100 |
| 56 | Ba | 131 | $\varepsilon\ \gamma$ | 80 |
| | Ba | 140 | $\beta^-\ \gamma$ | 4[f] |
| 57 | La | 140 | $\beta^-\ \gamma$ | 10 |
| 58 | Ce | 141 | $\beta^-\ \gamma$ | 40[f] |
| | Ce | 143 | $\beta^-\ \gamma$ | 10 |
| | Ce | 144 | $\alpha\ \beta^-\ \gamma$ | 5[f] |

TABLE 12.3 (*Continued*)

| Radionuclide | | | | MPBB in total |
|---|---|---|---|---|
| Z | | A | Decay[g] | body, $q$ (μc) |
| 59 | Pr | 142 | $\beta^- \gamma$ | 7 |
| | Pr | 143 | $\beta^-$ | 20 |
| 60 | Nd | 144 | $\alpha$ | 0.1[f] |
| | Nd | 147 | $\alpha \beta^- \gamma$ | 20 |
| | Nd | 149 | $\beta^- \gamma$ | 7 |
| 61 | Pm | 147 | $\alpha \beta^-$ | 60[f] |
| | Pm | 149 | $\beta^- \gamma$ | 20 |
| 62 | Sm | 147 | $\alpha$ | 0.1[f] |
| | Sm | 151 | $\beta^- \gamma$ | 100[f] |
| | Sm | 153 | $\beta^- \gamma$ | 30 |
| 63 | Eu | 152[d] | $\beta^- \varepsilon \gamma$ | 10 |
| | Eu | 152[e] | $\beta^- \varepsilon \gamma$ | 30 |
| | Eu | 154 | $\beta^- \varepsilon \gamma$ | 5[f] |
| | Eu | 155 | $\beta^- \gamma$ | 80[f] |
| 64 | Gd | 153 | $\varepsilon \gamma e^-$ | 90[f] |
| | Gd | 159 | $\beta^- \gamma$ | 20 |
| 65 | Tb | 160 | $\beta^- \gamma$ | 20[f] |
| 66 | Dy | 165 | $\beta^- \gamma$ | 10 |
| | Dy | 166 | $\beta^- \gamma e^-$ | 5 |
| 67 | Ho | 166 | $\beta^- \gamma e^-$ | 5 |
| 68 | Er | 169 | $\beta^- \gamma$ | 30 |
| | Er | 171 | $\beta^- \gamma e^-$ | 9 |
| 69 | Tm | 170 | $\beta^- \varepsilon \gamma e^-$ | 9[f] |
| | Tm | 171 | $\beta^-$ | 90[f] |
| 70 | Yb | 175 | $\beta^- \gamma$ | 30 |
| 71 | Lu | 177 | $\beta^- \gamma$ | 20 |
| 72 | Hf | 181 | $\beta^- \gamma$ | 100 |
| 73 | Ta | 182 | $\beta^- \gamma$ | 50 |
| 74 | W | 181 | $\varepsilon \gamma$ | 200 |
| | W | 185 | $\beta^-$ | 30 |
| | W | 187 | $\beta^- \gamma$ | 60 |
| 75 | Re | 183 | $\varepsilon \gamma$ | $2 \times 10^4$ |
| | Re | 186 | $\beta^- \gamma$ | 800 |
| | Re | 187 | $\beta^-$ | $4 \times 10^4$ |
| | Re | 188 | $\beta^- \gamma$ | 300 |
| 81 | Tl | 200 | $\varepsilon \gamma$ | $10^3$ |
| | Tl | 201 | $\varepsilon \gamma e^-$ | 400 |
| | Tl | 202 | $\varepsilon \gamma e^-$ | 200 |
| | Tl | 204 | $\beta^-$ | 100 |
| 82 | Pb | 203 | $\varepsilon \gamma$ | 400 |
| | Pb | 210 | $\alpha \beta^- \gamma$ | 0.7 |
| | Pb | 212 | $\alpha \beta^- \gamma e^-$ | 0.1 |
| 83 | Bi | 206 | $\varepsilon \gamma$ | 300 |
| | Bi | 207 | $\varepsilon \gamma$ | 300 |

TABLE 12.3 (*Continued*)

| Radionuclide | | | | |
|---|---|---|---|---|
| Z | | A | Decay[g] | MPBB in total body, $q$ (μc) |
| 83 | Bi | 210 | $\alpha\ \beta^-$ | 6 |
| | Bi | 212 | $\alpha\ \beta^-\ \gamma$ | 0.9 |
| 84 | Po | 210 | $\alpha$ | 0.5 |
| 88 | Ra | 223 | $\alpha\ \beta^-\ \gamma$ | 0.05[f] |
| | Ra | 224 | $\alpha\ \beta^-\ \gamma\ e^-$ | 0.06[f] |
| | Ra | 226 | $\alpha\ \beta^-\ \gamma$ | 0.1[f] |
| | Ra | 228 | $\alpha\ \beta^-\ \gamma\ e^-$ | 0.06[f] |
| 89 | Ac | 227 | $\alpha\ \beta^-\ \gamma$ | 0.03[f] |
| | Ac | 228 | $\alpha\ \beta^-\ \gamma\ e^-$ | 0.04[f] |
| 90 | Th | 227 | $\alpha\ \beta^-\ \gamma$ | 0.02[f] |
| | Th | 228 | $\alpha\ \beta^-\ \gamma\ e^-$ | 0.02[f] |
| | Th | 230 | $\alpha\ \gamma$ | 0.05[f] |
| | Th | 231 | $\alpha\ \beta^-\ \gamma$ | 30 |
| | Th | 232 | $\alpha\ \beta^-\ \gamma\ e^-$ | 0.04[f] |
| | Th | 234 | $\beta^-\ \gamma$ | 4[f] |
| | Th-natural | | $\alpha\ \beta^-\ \gamma\ e^-$ | 0.01[f] |
| 91 | Pa | 230 | $\alpha\ \beta^-\ \varepsilon\ \gamma$ | 0.07[f] |
| | Pa | 231 | $\alpha\ \beta^-\ \gamma$ | 0.02[f] |
| | Pa | 233 | $\beta^-\ \gamma$ | 60 |
| 92 | U | 230 | $\alpha\ \beta^-\ \gamma$ | $7 \times 10^{-3}$ |
| | U | 232 | $\alpha\ \beta^-\ \gamma\ e^-$ | 0.01[f] |
| | U | 233 | $\alpha\ \gamma$ | 0.05[f] |
| | U | 234 | $\alpha\ \gamma$ | 0.05[f] |
| | U | 235 | $\alpha\ \beta^-\ \gamma$ | 0.06[f] |
| | U | 236 | $\alpha\ \gamma$ | 0.06[f] |
| | U | 238 | $\alpha\ \gamma\ e^-$ | 0.06 |
| | U-natural | | $\alpha\ \beta^-\ \gamma\ e^-$ | 0.03 |
| 93 | Np | 237 | $\alpha\ \beta^-\ \gamma$ | 0.06[f] |
| | Np | 239 | $\alpha\ \beta^-\ \gamma$ | 30 |
| 94 | Pu | 238 | $\alpha\ \gamma$ | 0.04[f] |
| | Pu | 239 | $\alpha\ \gamma$ | 0.04[f] |
| | Pu | 240 | $\alpha\ \gamma$ | 0.04[f] |
| | Pu | 241 | $\alpha\ \beta^-\ \gamma$ | 0.9[f] |
| | Pu | 242 | $\alpha$ | 0.05[f] |
| 95 | Am | 241 | $\alpha\ \gamma$ | 0.05[f] |
| | Am | 243 | $\alpha\ \beta^-\ \gamma$ | 0.05[f] |
| 96 | Cm | 242 | $\alpha\ \gamma$ | 0.09 |
| | Cm | 243 | $\alpha\ \gamma$ | 0.09[f] |
| | Cm | 244 | $\alpha\ \gamma$ | 0.1[f] |
| | Cm | 245 | $\alpha\ \beta^-\ \gamma$ | 0.04[f] |
| | Cm | 246 | $\alpha$ | 0.05[f] |
| 97 | Bk | 249 | $\alpha\ \beta^-\ \gamma$ | 0.7[f] |
| 98 | Cf | 249 | $\alpha\ \gamma$ | 0.04[f] |
| | Cf | 250 | $\alpha$ | 0.04[f] |
| | Cf | 252 | $\alpha\ \gamma$ | 0.01[f] |

uniform distribution of radiation dose. However, despite these factors and with assumptions and restrictions, maximum permissible continuous exposure values can be estimated (Table 12.3). The values of body burden $q$ are based on that amount of the radionuclide which is deposited in the total body and produces the maximum permissible RBE dose rate to bone. The reader is urged to consult the report of the International Commission on Radiological Protection[2] committee for the complete tabulations of maximum permissible body burdens (MPBB), maximum permissible concentrations (MPC), and the methods for derivations. Only the values for the maximum permissible body burden are presented.

The effects of radiation from radioactive elements deposited within the skeleton are essentially similar to those of external radiation. Whatever the source of radiation, the general patterns of histological change are remarkably similar in different species. Histological damage includes empty lacunae, blood-vessel injury, irregular abnormal new bone, and varying degrees of fibrosis. In addition, in rats and mice where endochondral ossification continues in adult animals and in young rabbits, there may be unresorbed cartilage, abnormalities of cartilage in the epiphyseal plate, and severing of old trabeculae and abnormal resorption of new spongiosa.[6,7]

Bone damage takes two forms. First, bone may be injured indirectly through vascular injury and directly by action on the osteocytes. The presence of bone-seeking isotopes, particularly $Sr^{90}$ during a chronic phase of injury (180–200 days +), induces a sharp deterioration in the blood supply owing to the emptying of areas of the vascular bed of blood- and bone-forming tissues with the disruption of vascular innervation. Damage to osteocytes and vessels occurs within a few days in animals given a large short-term radiation, but it is most clearly seen as a late change, e.g., in bones of patients with radium poisoning. Second, radiation-damaged osteoblasts and osteoclasts can initiate abnormal activity in osteogenic connective tissue. Short-range alpha emitters affect the osteogenic connective tissue lining endosteal surfaces and resorption cavities of bone trabeculae and thus induce marked terminal fibrosis, especially when the dose is high. The longer-range beta emitters, $Sr^{90}$ and $P^{32}$, and external irradiation affect loose connective tissue in the bone-marrow spaces between bony trabeculae as well as on the surface of the trabeculae. These isotopes induce variable degrees of active cellular fibrosis often characterized by proliferation of pleomorphic spindle cells with conspicuous numbers of mitotic figures and abnormal giant cells. At higher dose levels, most changes are seen in different species. Their severity decreases with time, especially with an isotope of relatively short half-life (e.g., $P^{32}$) where irradiation is short compared with that of longer-lived isotopes. With a decrease in dose or termina-

tion of radiation, these changes become less severe, and, at sufficiently low doses, the initial damage is repaired so that no histological evidence of damage remains. Bone has a considerable capacity for repair.[1]

## 4. Radiation and Bone Tumors

A manifestation of delayed tissue damage produced by internally deposited radioisotopes, including radium, is malignancy; the induction period for appearance of the tumors following uptake of radium has varied as much as from 12 to 30 years. Of the malignant tumors of bone, the most frequent has been osteogenic sarcoma.

Information is accumulating on the body burden of radium in individuals who develop bone tumors after the long latent period following exposure. So far, no case has been reported in which less than 0.4 μc of radium,[8] plus an undetermined amount of mesothorium ($Ra^{228}$), was left in the body of an individual developing an osteogenic tumor who had been exposed to radium in adult life. On the other hand, one patient with a low terminal body burden of 0.8 μc developed a bone neoplasm.[9] Radium had been given medically 24 years previously, and the total accumulated dose from mesothorium, averaged over the bone, was estimated at 50 rads, plus 1100 rads from radium. These are dosage levels at which major skeletal damage in other patients has been observed.

The contribution of mesothorium to the radiotoxicity of luminous paints used by dial painters has been exhaustively reviewed by Aub *et al.*[10] While this contribution appears to have been appreciable in determining the end result of the toxicity of radioactive materials, it will be noted that in these instances $Ra^{228}$ was ingested as such. The effects are thus not comparable with those observed following the administration of thorotrast and the subsequent decay of $Th^{232}$ to $Ra^{228}$ (see Chapter 3).

The appearance of a bone tumor following radiation from an external source was first reported in 1922. Since then, many isolated cases have appeared.[9] In a report by Sabanas *et al.*,[11] 17 patients developed bone sarcoma, and 16 of these neoplasms appeared after irradiation for benign conditions. In another group of cases, osteosarcomas were observed in patients given radiation for reasons other than benign conditions of bone.

The effects of radioelements, other than radium, on bone have been rarely observed in man; most of our knowledge of these effects comes from studies on dogs, rabbits, rats, and mice. Comprehensive long-range studies on the toxicity of plutonium ($Pu^{239}$), radium ($Ra^{226}$), mesothorium ($Ra^{228}$), radiothorium ($Th^{228}$), and strontium ($Sr^{90}$) are under way at the Radiobiology Laboratory at the University of Utah. Osteogenic sarcomas have been pro-

duced in a colony of inbred beagles following single injections of the radioisotopes studied. Radium and strontium are found to be primarily deposited in areas where bone is being formed at the time of the injection, with some diffuse distribution. The isotopes of radium, e.g., $Ra^{228}$ (mesothorium), follow calcium in their distribution while plutonium, having no physiological analog, localizes on endosteal bone surfaces. The distribution of soluble thorium salts, e.g., $Th^{228}$ (radiothorium), is similar to that of plutonium. Osteogenic sarcomas, dose dependent and time dependent, have been observed to result from all these radioelements; the tumors so caused are frequently multiple. The relative biological effectiveness of the different isotopes depends mainly on the retained dose level of each.

Macpherson *et al.*[12] have reported on the relation of radiation dose to radiation damage in the tibia of weanling rabbits injected with strontium-90. Each of the rabbits was given a single injection of $Sr^{90}$ at two dose levels, 100 and 600 μc per kilogram, and the animals were then killed at various time intervals up to six months. Relatively little damage was seen at the lower dose level; at the higher level, there was gross damage. The earliest damage was to the cells in the region of maximum proliferation immediately below the epiphyseal plate, within one day after the injection. Disturbance of the blood vessels and epiphyseal cartilage plate was observed at 3 days. Areas of abnormal cellular proliferation, presumed to be presarcomatous lesions, were seen in osteogenic connective tissue at 3 months in regions which had received an accumulated dose of about 10,000 rads. Small discrete early tumors were sometimes noted in a region which had accumulated a similar dose. Multiple frank osteosarcomas had developed by 6 months. In similar experiments in weanling rats,[13] a severe impairment in the formation of the incisors was noted following administration of 100 μc of $P^{32}$.

Marshall and Finkel[7] have reported on comparison of microdosimetry and tumor production, following single injections of $Ca^{45}$, $Sr^{90}$-$Y^{90}$, or $Ra^{226}$ in mice. They conclude that hotspots produced by deposition of $Ca^{45}$ or $Ra^{226}$ do not deliver much of their dose to the region sensitive to the induction of osteogenic sarcoma and that the osteocytes and the volume of bone mineral are not the radiation-sensitive regions for osteogenic sarcomas. Instead, it is primarily the radiation escaping from microscopic bone surfaces that induces the tumor, a conclusion supported by the carcinogenic effectiveness of plutonium which remains at bone surfaces. It was suggested that the carcinogenic effectiveness of $Ca^{45}$ and $Ra^{226}$ hotspots is reduced by self-burial. These results are in accord with the conclusion of Vaughan[14] that osteogenic sarcomas induced by radiation arise at the surfaces of bone.

Strontium, like radium, is incorporated within the crystals of bone mineral, following an initial stage of rapid uptake by exchange for calcium ions at the

surface. The incidence of bone tumors in small animals depends on the dose of the radioelement. Single and multiple tumors and extensive metastases have been observed. Radioactive yttrium ($Y^{90}$) and cerium ($Ce^{144}$) also produce sarcomas in the skeletal system, especially in the long bones. Following sublethal doses of $P^{32}$, particularly in rats, a considerable proportion of the survivors develop malignant tumors, usually osteogenic sarcomas.

The pattern of bone-tumor response, as observed following introduction of $Sr^{89}$ or $Sr^{90}$ in $CF_1$ mice, is as follows: (1) the death rate from osteogenic sarcomas and other bone tumors, including multiple neoplasms, increases very rapidly after a long latent period; (2) a linear relation between dose and response has not been demonstrated; and (3) the data are consistent with the existence of a threshold below which sarcomas are not induced.[15]

## References


1. *Report of the United Nations Scientific Committee on the Effects of Atomic Radiation*, Suppl. No. 16 (A/5216), United Nations, New York, 1962.
2. International Commission on Radiological Protection, Permissible Dose for Internal Radiation, *Health Phys.*, **3**: 1-380 (1960).
3. R. D. Evans and N. C. Rasmussen, Isotopes: Radioactive Measurement, in *Medical Physics*, Vol. III, pp. 338-341, The Year Book Publ., Inc., Chicago, 1960.
4. R. A. Dudley, Natural and Artificial Radiation Background of Man, in *Low-Level Irradiation*, pp. 7-32, Am. Assoc. Advancement Sci., Washington, 1959.
5. M. Heller, Bone, in *Histopathology of Irradiation from External and Internal Sources*, pp. 70-161, McGraw-Hill Book Co., Inc., New York, 1948.
6. J. Vaughan, The Relation of Radiation Dose to Skeletal Damage from Bone-seeking Isotopes, *Sci. Basis Med. Ann. Rev.*, pp. 47-62 (1961).
7. J. H. Marshall and M. P. Finkel, Comparison of Microdosimetry and Tumour Production $Ca^{45}$, $Sr^{90}$, and $Ra^{226}$ in Mice, in *Radiation Damage in Bone*, pp. 40-44, Intern. Atomic Energy Agency, United Nations Document ST1/PUB/27, April 1960.
8. C. L. Dunham, Fallout from Nuclear Weapons Tests, *Advan. Biol. Med. Phys.*, **6**: 175-199 (1958).
9. L. L. Law, Radiation Carcinogenesis, *Advan. Biol. Med. Phys.*, **7**: 295-342 (1960).
10. J. C. Aub, R. D. Evans, L. H. Hempelmann, and H. S. Martland, The Late Effects of Internally-deposited Radioactive Materials in Man, *Medicine*, **31**: 221-329 (1952).
11. A. O. Sabanas, D. C. Dahlin, D. S. Childs, Jr., and J. C. Ivins, Postradiation Sarcoma of Bone, *Cancer*, **9**: 528-542 (1956).
12. S. Macpherson, M. Owen, and J. Vaughan, The Relation of Radiation Damage to Radiation Dose from Strontium 90, *J. Bone Joint Surg.*, **42B**: 395 only (1960).
13. G. C. H. Bauer, A. Carlsson, and B. Lindquist, Impairment in Incisor Teeth Formation Following Administration of High Activity $P^{32}$ in Weanling Rats, *Acta Radiol.*, **48**: 97-100 (1957).
14. J. Vaughan, The Relation of Radiation Dose to Radiation Injury in the Skeleton, with Special Reference to $^{90}Sr$, in *Radioisotopes and Bone*, pp. 317-333, Blackwell Scientific Publ., Oxford, 1962.
15. M. P. Finkel, P. J. Bergstrand and B. O. Biskis, The Consequences of the Continuous Ingestion of $Sr^{90}$ by Mice, *Radiology*, **74**: 458-467 (1960).

# Chapter 13

# *Natural Radioactivity and Radioactive Fallout*

## 1. Natural Radioactivity

The sources of natural radiation are cosmic rays (of external origin) and naturally occurring radioactive elements, which together give rise to both the external and internal radiation of the human body. The distinguishing feature of irradiation by natural sources is that the entire population of the world has been exposed to it since the beginning and will continue to be exposed to it in the future.[1] We are not here concerned with the phenomena of cosmic radiation but only with its contribution in dose-rate to the human body.

Cosmic radiation consists of primary particles and secondary radiation. The sources of primary particles are galactic cosmic radiation, geomagnetically trapped radiation, and solar cosmic radiation. Primary particles are composed of about 80% protons (hydrogen nuclei), 19% alpha particles (helium nuclei), and 1% heavier elements, e.g., O, N, C, Ca, and Fe. Each particle is completely ionized, stripped of its electrons, and carries a positive charge equal to its atomic ($Z$) number.[2] Primary particles interact with nuclei of the upper atmosphere, producing secondary radiation.

When primary particles interact with nuclei of the upper atmosphere, there is produced secondary radiation consisting of a soft component (i.e., electrons and photons) and a nucleonic component (i.e., protons, mesons and neutrons). Secondary radiation is a fairly steady stream of high-energy particles produced from the explosive collision of primary cosmic rays with atmospheric nuclei; these collisions in turn produce streams of secondary and tertiary particles. The gamma rays formed are usually all that is observed at the earth's surface; this gamma radiation contributes to background radiation.[3]

Very few primary particles reach the earth's surface owing to either dissipation of total energy by ionization and excitation or by collision with atmospheric nuclei. There is an increase in cosmic radiation dose rate with altitude; above 60,000 feet it is a hazard owing to high specific ionization of the heavy nuclei, and it becomes the major dose-rate contributor to man. In the altitude region from about 60,000 to 120,000 feet, secondary radiation decreases; above this altitude region, the radiation dose-rate is predominantly from primary particles. It is suggested that 50 mrem per year be taken as a

typical value for the total cosmic radiation tissue dose-rate for sea level at middle latitudes. The exposure nearly doubles for each 3,000 feet increase in altitude.[1]

## 2. Naturally Occurring Radioactive Elements

Naturally occurring radionuclides are widely distributed throughout the earth's crust. Natural radioactivity also exists in various sources of water and in the lower atmosphere; eventually the radioactivity is transferred to man via the food chain from soil and water.

The gamma-ray dose to man from the naturally occurring radionuclides is largely governed by their abundance in nature. The important radionuclides are the long-lived $U^{238}$, $Th^{232}$, and $Ra^{226}$ and their daughter products; $K^{40}$; and the cosmic ray-produced $C^{14}$ and $H^{3}$. Other radionuclides (such as $Rb^{87}$, $La^{138}$, $Sm^{147}$, and $Lu^{176}$) do not contribute significantly to the dose received by man owing to their scarcity. Since uranium and thorium are almost universally distributed, their daughter products which accumulate in soil and rocks will contribute over half the observed gamma activity of uranium minerals. The beta-particle emission of these radionuclides can contribute an external dose to skin but the dose-rate delivered to critical organs such as gonads and bone is regarded as negligible.[1] Table 13.1 lists some of the radionuclide contents of various rocks. The importance of these values is reflected in terrestrial dose-rate measurements where the radioactivity of building materials must be taken into consideration in the estimation of mean dose-rate to a population.[1]

The levels of radioactivity in water originating from the uranium-radium and thorium series (see Chapter 3) are high in certain natural springs in areas where there are high concentrations of uranium and thorium in the soil. Typical concentrations of radioactivity in sea water are 0.2–0.9 μμc of $U^{238}$ per liter, 0.001–0.1 μμc of $Th^{232}$ per liter, 0.02–0.3 μμc of $Ra^{226}$ per liter, and 300 μμc of $K^{40}$ per liter. Spring water contains high concentrations of $Rn^{222}$; however, the resulting mean dose to the world population is negligible owing to the short half-life of the radioisotope. Public water supplies may contain $Ra^{226}$, $Rn^{222}$, and $Th^{232}$ and their decay products.[1] Some of the radioactivity can be effectively removed from water supplies by ion exchange.[4]

Only about 10% of the $Ra^{226}$ enters the body through water and about 90% from food. When the level of $Ra^{226}$ in water is high and exceeds the intake from food, some correlation is found between the skeletal concentration and the concentration in water. With relatively low $Ra^{226}$ concentration in water, no such correlation is observed since the skeletal $Ra^{226}$ content is then determined by the intake from food.[4]

There is a wide range of radioactivity in vegetation; however, there is no simple correlation between it and the radioactivity in soil, for which the range is smaller. Since there is not much information about discrimination factors for the soil-food and food-man chains, the dose-rates to particular body organs are best related to measured concentrations in the organs.[1]

TABLE 13.1
MEAN VALUES ($\mu\mu c/g$) OF RADIONUCLIDE CONTENTS IN VARIOUS ROCKS[a]

| Type of rock | $U^{238}$ | $Th^{232}$ | $Ra^{226}$ | $K^{40}$ |
|---|---|---|---|---|
| Igneous rocks | 0.5 | — | — | — |
| | 1.4 | 1.3 | 1.3 | 22 |
| | — | 3.5 | 3.7 | 25 |
| Granites | 1.4 | — | — | — |
| | — | — | 2.6 | — |
| | 3 | 13 | — | — |
| Basalts | — | 1.0 | 1.1 | 11 |
| | — | — | — | 8 |
| | — | — | 0.5 | — |
| Volcanic rocks: | | | | |
| Basic lavas | 0.9 | 3 | — | — |
| Acidic lavas | 5 | 15 | — | — |
| Sedimentary rocks: | — | — | 0.25 | — |
| Sandstones | 0.4 | 0.7 | 0.7 | 9 |
| Sandstones | — | — | 0.3 | — |
| Limestone | 0.5 | 0.15 | 0.4 | 2.3 |
| Limestone | — | 0.1 | 0.7 | 2.5 |
| Limestone | — | — | 1 | 2.5 |
| Alum shales (Sweden) | — | 0.1 | 60 | — |
| Alum shales (Sweden) | — | 0.17 | 60 | 29 |
| Shales | 0.5 | 0.14 | 0.4 | 2.3 |
| Bituminous shale (U.S.A.) | 22 | — | — | — |
| Phosphate rocks (Florida) | 40 | — | — | — |
| Phosphate rocks (Nauru) | 20 | 0.8 | — | — |
| Phosphate rocks (N. Africa) | 4 | 1.0 | — | — |
| Phosphate rocks (N. Africa) | 8 | — | — | — |

[a] From *Report of UN Sci. Comm. on Effects of Atomic Radiation.*[1]

In some areas of the world, there is a high natural radiation background from the cosmic ray altitude effect and also from high concentrations of radioactive materials in the soil, drinking water, air, and general environment. Data from these areas are shown in Table 13.2. Studies of the population in these areas might give valuable information concerning genetic risks accumulating from the increased exposure to ionizing radiation. There is also the possibility of additional radiation dose to the populations in these areas

TABLE 13.2
SPECIAL AREAS OF HIGH EXTERNAL RADIATION FROM RADIOACTIVITY IN SOIL OR ROCK[a]

| Area | Approximate population | External dose-rate in air (cosmic plus terrestrial) | Measuring instruments |
|---|---|---|---|
| Monazite area in states of Rio de Janeiro and Espirito Santo, Brazil; sequence of coastal strips, each several miles long and several hundred feet wide | 30,000 | Average 500 mrad/year, peak 1000 mrad/year | Ionization chamber, scintillation counter |
| Mineralized volcanic intrusives in states of Minas Gerais and Goias, Brazil; 3.6 miles in a dozen scattered places | 1 village with 350 inhabitants, pasture land and scattered farms | Average 1600 mrad/year, peak 12,000 mrad/year | Ionization chamber, scintillation counter |
| Primitive granitic, schistous and sandstone areas of France; area includes about 1/6 of French population | 7,000,000 | 180–350 mrad/year | Geiger counter, scintillation counter (NaI and plastic phosphor) |
| Monazite area in Kerala and Madras states, India; approximately 120 miles long and several hundred feet wide | 100,000 | Population weighted mean 1300 mrad/year plus about 200 mrad/year beta rays | Ionization chamber |
| Niue Island, Pacific; volcanic soil and unusually high radioactive content of plants | 4,500 | Maximum external radiation 1000 mrad/year. High total alpha radioactivity content 50–360 $\mu\mu c/g$ in the main foodstuff—taro | — |
| Monazite areas of Northern Nile Delta region, UAR | "Highly populated" | 300 and 400 mrad/year in 2 villages; 4 other villages in same area 110–150 mrad/year | Scintillation counter |

TABLE 13.3
TOTAL BODY $Ra^{226}$ CONTENTS[a]

| Sampling locality | Number of samples | Concentration $Ra^{226}$ in tap water, μμc/liter | $Ra^{226}$ body content μμc Range | Mean | Type of sample |
|---|---|---|---|---|---|
| U. S. A., Rochester, N. Y. | 20[b] | 0.04 | 38–353 | 118 | Whole body ash |
| U. S. A., Rochester, N. Y. | 14[b] | 0.04 | 47–130 | 87 | Whole body ash |
| U. S. A., Rochester, N. Y. | 9 | 0.04 | 15–65 | 30 | Body organ ash |
| U. S. A., Northwest Pacific area | 50 | 0.001 | 13–139 | 47 | Whole body ash |
| Germany, Frankfurt-on-Main | 15 | 0.14–0.31 | 130–790 | 330 | Body organ ash |
| Germany, Frankfurt-on-Main | Up to 56 | 0.2 | — | 130 (35 in skeleton) | Body organ ash |
| U. S. A., New York | 140 | 0.04 | 3–150 | 24 | Whole skeleton ash |
| Six different countries | 21 | — | — | 40 | Single bone ash |
| Thirteen different countries | 499 | — | — | 33 | Bone ash; composite sample |
| U. S. A., prisoners,[c] 4 months detention | 11 | 3.4 | — | 100 | *In vivo*, radon in breath |
| U. S. A., prisoners,[c] 7.6 years detention | 8 | 3.4 | — | 202 | *In vivo*, radon in breath |
| U. S. A., prisoners,[c] 19.7 years detention | 11 | 3.4 | — | 236 | *In vivo*, radon in breath |
| U. S. A., Lockport boys[c] | 8 | 8 | — | 368 | *In vivo*, radon in breath |
| U. S. A., Chicago boys | 7 | 0.03 | — | 36 | *In vivo*, radon in breath |
| U. S. A., low-activity water | 42 | < 0.1 | 15–81 | 36 | Single bone ash |
| U. S. A., high-activity water | 34 | 0.1–10.5 | 36–1400 | — | Single bone ash |

[a] From *Report of UN Sci. Comm. on Effects of Atomic Radiation*.[1]
[b] Check series using different methods of measurement.
[c] From Joliet, Illinois, area.

from internal deposition of radioactive material. As yet no published data are available on this aspect.[1]

Natural radioactivity in man is present at very low levels; methods used for the detection and measurement of these low concentrations of radioactivity have been discussed in Chapter 10. This section is concerned with dose rates obtained by various methods. The important naturally occurring radioisotopes in man are those of the uranium-radium and thorium series,

TABLE 13.4
BODY TISSUE DOSE-RATES DUE TO EXTERNAL AND INTERNAL IRRADIATION FROM NATURAL SOURCES OF RADIATION IN "NORMAL" REGIONS[a]

| Source of irradiation | Dose-rate in mrem/year | | |
|---|---|---|---|
| | Gonad | Haversian canal | Bone marrow |
| *External irradiation:* | | | |
| Cosmic rays (including neutrons) | 50 | 50 | 50 |
| Terrestrial radiation (including air) | 50 | 50 | 50 |
| *Internal irradiation:* | | | |
| $K^{40}$ | 20 | 15 | 15 |
| $Ra^{226}$ and decay products (35% equilibrium) | 0.5 | 5.4 | 0.6 |
| $Ra^{228}$ and decay products (equilibrium) | 0.8 | 8.6 | 1.0 |
| $Pb^{210}$ and decay products[b] (50% equilibrium) | 0.3 | 3.6 | 0.4 |
| $C^{14}$ | 0.7 | 1.6 | 1.6 |
| $Rn^{222}$ (absorbed into blood stream) | 3 | 3 | 3 |
| | 125 | 137 | 122 |

[a] From *Report of UN Sci. Comm. on Effects of Atomic Radiation.*[1]
[b] $Pb^{210}$ in excess of that expected from $Ra^{226}$ and decay products in 35% equilibrium.

radiopotassium, radiocarbon, and tritium. Measurement of the body potassium content is done by whole-body counting of gamma rays. Total radium content is usually estimated by measuring the rate of exhalation of radon; a general value is assumed for the fraction of the total radon formed which escapes in the breath. Another method used is the analysis of autopsy samples. In Table 13.3 are listed values for total body $Ra^{226}$ content obtained by different methods of analysis.

Body tissue dose-rates from natural sources of radioactivity are presented in Table 13.4. The value for total cosmic radiation tissue dose-rate, 50 mrem per year, is the value at sea level for middle latitudes. The terrestrial gamma radiation results from gamma activities present in the soil, buildings, and

air, and the average tissue dose-rate of 50 mrem per year takes into account absorption of radiation by outer tissues. The dose-rates from tritium are very small compared with those from the other radionuclides listed.

## 3. Radioactive Fallout

Nuclear fission occurs both in nuclear reactors and in the instantaneous detonation of a nuclear device. In both instances cleavage of a uranium or a plutonium nucleus leads to the formation of two atomic nuclei, each of

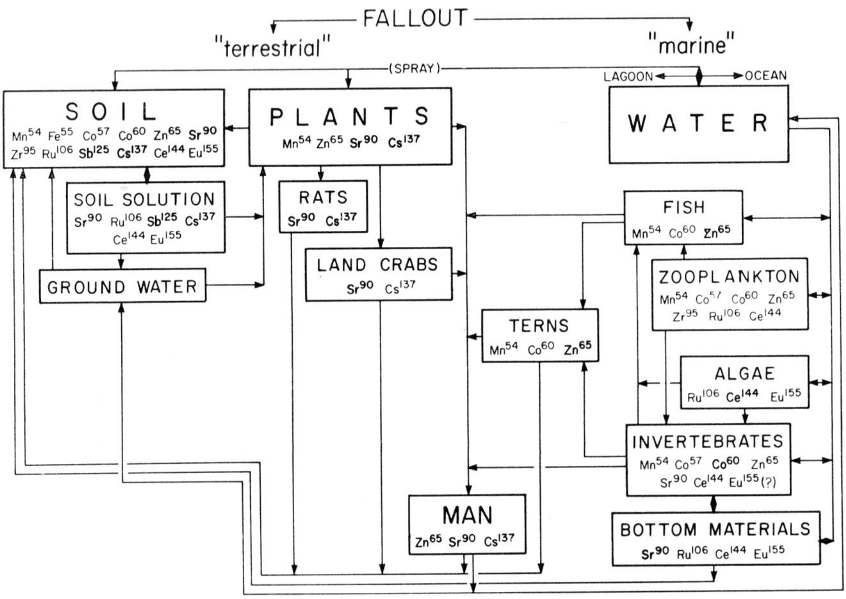


FIG. 13.1. Distribution of fallout radioisotopes. (From Seymour, *Applications of Radioisotopes and Radiation in the Life Sciences.*[5])

which is radioactive owing to its instability. These fission products include radioactive isotopes of the 34 elements in the periodic system, from zinc (element 30) to the rare earth europium (element 63). In some instances, several isotopes of the same element may be formed as fission products; in all, a total of some 200 radioactive substances are present in fallout. The mixture of fission products emits both beta particles and penetrating gamma rays. It includes elements whose half-lives vary from a fraction of a second to many years; the intensity of the radiation falls off sharply with time owing to the rapid decay of the elements with short half-lives (Fig. 13.1).

According to the circumstances under which a nuclear device is detonated, fallout may be either local or world-wide or a mixture of the two. Local fall-

out occurs when the detonation is close enough to the surface of the earth to bring large quantities of soil or water into the fireball where molten particles can condense onto this material. To date, with the tragic exceptions of Hiroshima and Nagasaki, man has not been exposed to local fallout except for the accidents which involved the Marshall Islanders[6] and the 23 Japanese fishermen.[7] When the airburst is sufficiently above the earth's surface, so that dirt and water are not drawn into the fireball, fission products are disseminated either into the troposphere or into the stratosphere. Tropospheric or intermediate fallout is ordinarily deposited on the earth's surfaces within 30 to 60 days; stratospheric or delayed fallout may take many months or years for deposition.

TABLE 13.5
RETENTION OF FISSION PRODUCTS IN BONE[a]

| No. (*Z*) | Element | Symbol | Mass no. | Type of radiation | Days retained in |
|---|---|---|---|---|---|
| 38 | Strontium | Sr | 89, 90 | $\beta^-$ | Bone > 200 |
| 39 | Yttrium | Y | 91 | $\beta^-, \gamma$ | Bone > 500 |
| 40 | Zirconium | Zr | 95 | $\beta^-, \gamma$ | Bone > 100 |
| 41 | Niobium | Nb | 95 | $\beta^-, \gamma$ | Bone > 30 |
| 44 | Ruthenium | Ru | 103, 106 | — | — |
| 52 | Tellurium | Te | 127, 129 | — | — |
| 53 | Iodine | I | 131 | — | — |
| 54 | Xenon | Xe | 133 | — | — |
| 55 | Cesium | Cs | 137 | — | — |
| 56 | Barium | Ba | 140 | $\beta^-, \gamma$ | Bone > 50 |
| 57 | Lanthanum | La | 140 | $\beta^-, \gamma$ | Bone > 25 |
| 58 | Cerium | Ce | 144 | $\beta^-, \gamma$ | Bone > 200 |
| 59 | Praseodymium | Pr | 143 | $\beta^-$ | Bone > 100 |
| 61 | Promethium | Pm | 147 | $\beta^-$ | Bone > 100 |

[a] From McLean and Urist, *Bone: An Introduction to the Physiology of Skeletal Tissues.*[8]

Among the 34 fission-product elements, there are 14 whose rates of decay are expressed in terms of days to years and which are produced in relatively large amounts by fission. Almost all of the radioactivity remaining in a fission-product mixture after it has decayed for a week comes from the isotopes of these 14 elements (Table 13.5). Of these, nine isotopes are stored in bone for periods of 25 to 500 days or more and, for this reason, constitute a hazard to man if they are absorbed and reach the circulation. In so far as the skeleton is concerned, the hazards from absorption and deposition of these radioactive elements include (1) damage to the function of the bone marrow and depletion of its formed elements, (2) initiation of malignant tumors, and (3) direct damage to bone structure and interference with normal growth and repair.

Of the radioisotopes present immediately following detonation of a nuclear device, most have short physical half-lives and decay in a short time or have other characteristics, such as insolubility, that result in little or no hazard to man. Those that warrant special consideration from the point of view of hazard may be limited to a total of four: $I^{131}$, $Cs^{137}$, $C^{14}$, and $Sr^{90}$. Iodine-131 has no special affinity for bone and will not be considered here. Cesium-137 is of theoretical interest in connection with bone, but it does not constitute a special hazard to the skeletal system (see Chapter 3). Carbon is an integral part of all living cells and of organic matter; in addition, it may be deposited in bone as $C^{14}O_3^{2-}$. Although it has been demonstrated in the bone of experimental animals, it has not been shown to lead to skeletal damage. Of the nine radioisotopes stored in bone (Table 13.5), and potentially hazardous for that reason, only strontium is: (1) found in sufficient amounts among the fission products, (2) is readily enough absorbed by the gastrointestinal tract, and (3) has a sufficiently long half-life to be considered a hazard by reason of its radiation effects upon bone tissue. Of the radioisotopes of strontium liberated by nuclear fission, only $Sr^{90}$ has a long half-life (27.7 years); that of $Sr^{89}$ is 51 days. Thus, of all the fission products, the long-term hazard by virtue of quantity, uptake, and retention in bone narrows down to the one radioisotope, $Sr^{90}$.

## 4. Strontium-90

Most of the $Sr^{90}$ found in man and animals as a result of contamination of the atmosphere gains entrance by ingestion in foods; the pathway is from air to plants, via soil or direct uptake by leaves, and from plants to man, either direct or via the milk of animals ingesting contaminated plants. The amount directly absorbed into the human organism by inhalation is small. The physiological behavior of strontium in the body, as contrasted with calcium, is considered in Chapter 8, as is the radiotoxicity resulting from radiostrontium deposited in the bones. In this section attention will be directed to the over-all problem of the fallout hazard, especially that from $Sr^{90}$.

Strontium-90 may become associated with foodstuffs by surface contamination (as when fallout settles on the exposed surfaces) or by uptake from soil. During a period of relatively heavy fallout, as in 1959, surface contamination may account for as much as 80% of the activity found in plants. In subsequent periods, although the relative uptake from the soil becomes more important, the total contamination of the plants is less. The over-all hazard is thus in direct proportion, not necessarily linear, to the total amount of fallout. Fallout of $Sr^{90}$ originates chiefly from the stratosphere; its greatest deposition rates are in the spring and early summer months and in areas of heavy rainfall.

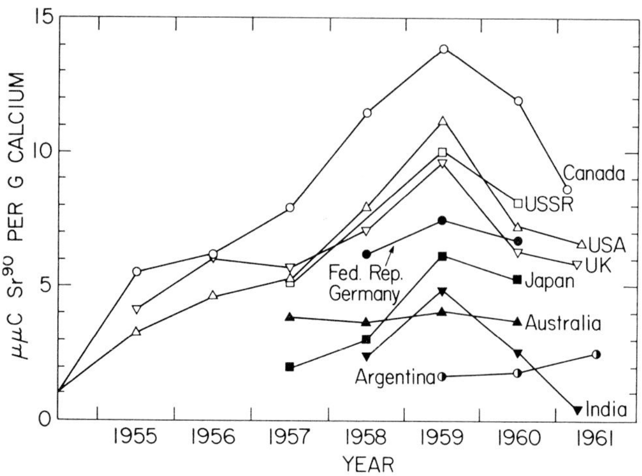


FIG. 13.2. Yearly mean concentrations of $Sr^{90}$ per gram of calcium in milk from nine countries. (From *Report of the UN Sci. Comm. on Effects of Atomic Radiation.*[1])

TABLE 13.6
ESTIMATES OF ANNUAL INTAKE OF CALCIUM AND STRONTIUM-90[a]

| Food category | Calcium (g/year) | Strontium-90 (μμc) Chicago, July 1961 | New York, May 1961 | San Francisco, June 1961 |
|---|---|---|---|---|
| Bakery products | 37.0 | 229 | 381 | 79 |
| Whole grain products | 10.0 | 87 | 86 | 56 |
| Eggs | 9.1 | 35 | 27 | 22 |
| Fresh vegetables | 15.0 | 103 | 490 | 47 |
| Root vegetables | 6.1 | 53 | 428 | 46 |
| Milk | 234.3 | 995 | 1967 | 420 |
| Poultry | 9.2 | 65 | 14 | 13 |
| Fresh fish | 10.8 | 3 | 5 | 6 |
| Flour | 8.6 | 258 | 232 | 120 |
| Macaroni | 0.7 | 11 | 11 | 9 |
| Rice | 1.1 | 4 | 4 | 3 |
| Meat | 10.9 | 58 | 44 | 45 |
| Shellfish | 0.8 | 1 | 1 | 3 |
| Dried beans | 2.9 | 46 | 50 | 23 |
| Fresh fruit | 13.6 | 48 | 279 | 81 |
| Potatoes | 5.8 | 41 | 63 | 57 |
| Canned fruit | 1.3 | 23 | 26 | 33 |
| Fruit juices | 1.7 | 42 | 44 | 30 |
| Canned vegetables | 4.2 | 80 | 62 | 33 |
| Total annual intake | 383 | 2182 | 4214 | 1129 |

[a] From Dunning, *Borden's Rev. Nutr. Res.*[9]

These phenomena are reflected in the $Sr^{90}$ found in milk, as illustrated in Fig. 13.2.

Table 13.6 gives estimates of the annual intake of $Sr^{90}$ and of stable calcium from a variety of foodstuffs as calculated from observations in three principal cities in the United States—Chicago, New York, and San Francisco. According to these data, 30 to 50% of the $Sr^{90}$ intake in the food may come from milk and milk products. Other important sources of the radioisotopes may be fresh fruit and vegetables, as well as flour and bakery products. The relative amounts from different sources will vary widely with the time and place of sampling; milk, however, may be expected to constitute the chief source under nearly all circumstances, with the dietary habits common in the United States. The amounts of $Sr^{90}$ in the atmosphere, and hence the amounts absorbed by man, have varied directly with the amounts released by the testing of nuclear weapons; there was a marked decrease in the dissemina-

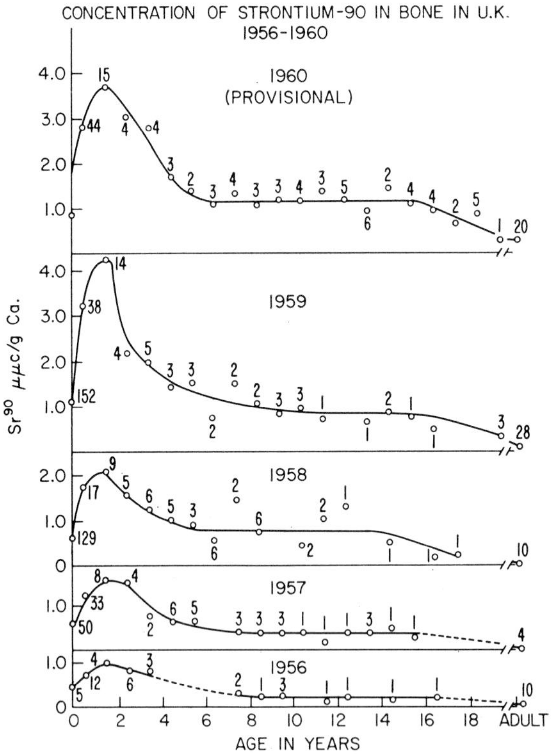


FIG. 13.3. The mean specific activity of bone (femur) as micromicrocuries of strontium-90 per gram of calcium according to age in the years 1956–1960. Data from United Kingdom. (From Loutit, *Irradiation of Mice and Men*.[11])

tion of $Sr^{90}$ during the moratorium on nuclear tests from November 1958 to August 1961.[10]

### 4.1. Strontium-90 in Bone

The Lamont Geological Observatory, Columbia University, Palisades, New York, has carried out world-wide observations on the concentrations of $Sr^{90}$ in human bone since 1953. During this period, more than 10,000 samples of bone were received from a network of 39 stations.[10] From these data, it has been possible to predict the $Sr^{90}$ concentration in the world population, under given conditions, with moderate reliability. Radiation doses to the skeleton to be expected from world-wide fallout on the basis of a 3000-megaton (fission yield) war are also given. It is not within the scope of this chapter to review all of the findings of these investigators; their results, however, afford an excellent example of the potentialities of isotopic methods of analysis.

Figure 13.3 illustrates the concentration of $Sr^{90}$ in human bone in the United Kingdom[11] for the years 1956–1960 inclusive. The striking features of these data are (1) the high values in young children, increasing progressively from 1956 to 1959–1960, (2) the approximately constant values for adults, and (3) the rising values in the newborn. As Kulp and Schulert[10] have pointed out, the fetus reflects the $Sr^{90}$ concentration in the human being more rapidly than any other age group, since most of the skeletal structure is produced during the last 3 months of pregnancy. The $Sr^{90}$ content of human bones varies directly with the amounts in the atmosphere and in milk, but reflects a cumulative dose rather than the immediate intake.

### 4.2. Strontium-90 and Milk

As discussed above, the chief source of $Sr^{90}$ for man, as a consequence of fallout, is dairy milk. In addition to these data, it has been shown by Carr *et al.*[12] that the availability of strontium in milk is equal to that of strontium as chloride in dilute aqueous solution. These findings have led to some apprehension concerning milk as a foodstuff, particularly for young growing children. This has manifested itself in two ways: (1) by a decrease in the consumption of milk, and (2) by attempts to devise means for the removal of $Sr^{90}$ from milk.

As of February 13, 1962, the U.S. Public Health Service issued a statement entitled "Ion Exchange Process for Removing Radioactive Strontium-90 from Milk" in which it is reported that removal of up to 90% of radioactive strontium has proved practicable on a pilot plant scale, although its practicability on a commercial scale has not yet been tested. Even more important is the statement that strontium-90 and strontium-89 levels over the next year will be well below the daily intake, averaged over a year's time, recommended by

the Federal Radiation Council as the point at which consideration need be given to possible use of countermeasures designed to reduce the intake of radioactivity. It is stated further that, since milk is an essential source of calcium for the bones and teeth of growing children, any widespread disruption of its production and distribution is likely to create individual and public health hazards far more widespread and serious than will be created by the small amounts of radioactive exposure involved at the levels of strontium-89 and strontium-90 in the years ahead.[9]

## 5. Removal of Radioisotopes from Bone

Many attempts have been made to find methods by which radioactive elements, once absorbed into the body, can be prevented from depositing in bone or, having been so deposited, can be removed from the bone and from the organism. Two types of approach are under current consideration: (1) the use of carrier agents, such as zirconium, which form colloidal aggregates with the radioelement in the blood and which then are removed from the circulation by excretion through the kidneys; and (2) complexing or chelating agents which form soluble, nonionic, readily excreted complexes with the radioisotope. In applications of the carrier therapy zirconium has been used successfully in experimental animals to remove plutonium, yttrium, and cerium; it is more effective if given as soon as possible after the radioisotope enters the blood. Zirconium has not been useful in increasing elimination of strontium or of thorium.

The chelating agents, especially the polyamino compounds such as ethylenediamine tetraacetic acid (EDTA) and diethyltriamine pentaacetic acid (DTPA), have proved useful for removing a number of isotopes from the bodies of experimental animals, including plutonium, thorium, yttrium, and the rare earths, such as cerium. Again, these agents have not been effective against strontium or radium.[13]

In no case can one expect that the removal of an element from the stable bone mineral will be practicable, since this would require virtual demineralization of the bone. The above methods may be useful only for the isotopes still in the blood or, at the most, those in the readily exchangeable portions of the bone mineral.

## 6. Whole-Body Exposure

Various estimates of the whole-body exposure, i.e., external radiation plus internal $Cs^{137}$ and $C^{14}$ from past nuclear-weapons tests have been made.[1] Table 13.7 lists the dose commitment from past and future nuclear testing. These estimates generally range around 100 milliroentgens over a 30-year

period from the time of exposure and represent only a fraction of the natural background radiation to be expected over that period. According to the estimates of Kulp and Schulert,[10] in the event of a nuclear war the gamma

TABLE 13.7

Dose Commitment from Nuclear Testing[a]

| Tissue or organ | Source of radiation | Tests 1954–1961: Dose commitment (mrem) | Tests 1954–1961: Fraction of dose commitment reached by year 2000 | Future tests dose commitment per year of testing (mrem) |
|---|---|---|---|---|
| Gonads | External | 30 | 0.97 | 3.8 |
| | Internal: | | | |
| | $Cs^{137}$ | 11 | 1.0 | 3.1 |
| | $C^{14}$ | 70 | 0.10 | 22 |
| Cells lining bone surfaces | External | 30 | 0.97 | 3.8 |
| | Internal: | | | |
| | $Sr^{90}$ | 79 | 0.91 | 10.5 |
| | $Cs^{137}$ | 19 | 1.0 | 5.3 |
| | $C^{14}$ | 116 | 0.10 | 37 |
| Bone marrow | External | 30 | 0.97 | 3.8 |
| | Internal: | | | |
| | $Sr^{90}$ | 40 | 0.91 | 5.3 |
| | $Cs^{137}$ | 14 | 1.0 | 3.9 |
| | $C^{14}$ | 70 | 0.10 | 22 |

[a] From *Report of UN Sci. Comm. on Effects of Atomic Radiation.*[1]

flux from the shorter-lived isotopes would dominate the total dose to bone during the first year after the event, but $Sr^{90}$ would be the primary source of radioactivity thereafter.

## References

1. *Report of the United Nations Scientific Committee on the Effects of Atomic Radiation,* Suppl. No. 16 (A/5216), United Nations, New York, 1962.
2. D. G. Simons, Biologic Hazard of Space Radiations, in *Medical Physics,* Vol. III, pp. 64-72, The Year Book Publishers, Inc., Chicago, 1960.
3. R. T. Overman and H. M. Clark, *Radioisotope Techniques,* McGraw-Hill Book Company, Inc., New York, 1960.
4. H. F. Lucas, Jr., *Correlation of the Natural Radioactivity of the Human Body to That of Its Environment: Uptake and Retention of Ra-226 from Food and Water,* USAEC Report ANL-6297, pp. 55-56, Argonne National Laboratory, 1961.
5. A. Seymour, *Statement in Hearings before the Subcommittee on Research, Development, and Radiation, Applications of Radioisotopes and Radiation in the Life Sciences,* pp. 266-271, U. S. Government Printing Office, Washington, D. C., 1961.
6. S. H. Cohn, J. S. Robertson, and R. A. Conard, Radioisotopes and Environmental

Circumstances: The Internal Radioactive Contamination of a Pacific Island Community Exposed to Local Fallout, in *Radioisotopes in the Biosphere*, pp. 306-330, University of Minnesota, Minneapolis, 1960.
7. M. Tsuzuki, Radiation Injury Due to Radioactive Fallout, in *Proc. Intern. Conf. Peaceful Uses of Atomic Energy*, Vol. II, pp. 132-133, United Nations Publications, New York, 1956.
8. F. C. McLean and M. R. Urist, *Bone: An Introduction to the Physiology of Skeletal Tissue*, University of Chicago Press, Chicago, 1955.
9. G. M. Dunning, Foods and Fallout, *Borden's Rev. Nutr. Res.*, **23**: 1-15 (1962).
10. J. L. Kulp and A. R. Schulert, Strontium-90 in Man V, *Science*, **136**: 619-632 (1962).
11. J. F. Loutit, *Irradiation of Mice and Men*, University of Chicago Press, Chicago, 1962.
12. T. E. F. Carr, G. E. Harrison, and A. Sutton, Availability to Man of Strontium in Cow's Milk, *Proc. Soc. Exptl. Biol. Med.*, **110**: 151-154 (1962).
13. M. W. Rosenthal, Radioisotope Absorption and Methods of Elimination: Factors Influencing Elimination from the Body, in *Radioisotopes in the Biosphere*, pp. 541-563, University of Minnesota, Minneapolis, 1960.

# Chapter 14

# *The Pathology and Pathological Physiology of Bone*

## 1. Introductory Statement

In the foregoing chapters, attention has been directed toward the biochemistry and physiology of bone, particularly as elucidated with the aid of radioisotopes. This chapter deals with the application of isotopes to the pathology and pathological physiology of bone, including special reference to their use in the clinic, and gives consideration to both localized lesions in bone and metabolic disorders of the skeleton.

For clinical purposes, the methods generally require administration of an isotope to the patient, usually $Ca^{45}$, $Ca^{47}$, or $Sr^{85}$ in a single dose given intravenously. Following this, there is a variety of methods for observation, including blood-disappearance curves, excretion in the urine and feces, and external counting (either by total-body counts or by scanning over the bones). The kind of information given by each of these methods has been considered in earlier chapters; the usefulness of these methods, especially those most commonly employed, will constitute the subject matter of this chapter.

## 2. Kinetic Methods

Application of kinetic methods to the study of skeletal metabolism is usually directed toward the determination of two parameters: (1) the $A$ value of Bauer *et al.*,[1] which corresponds to the bone formation rate ($BFR$) of Heaney and Whedon[2] (stated in terms of grams of calcium per day, for the whole body, or per unit weight of the body, e.g., per kilogram per day); and (2) the $E$ value of Bauer *et al.*,[1] which corresponds to the miscible calcium pool of Heaney and Whedon[2] (computed in terms of grams of calcium).

The principles utilized in arriving at values for $A$ and $E$ are outlined in Chapter 7. Complete data are needed from two separate time intervals in addition to determinations of the serum activity at least every 24 hours. Each group of workers, while following essentially the same principles, has modified the procedures and also, to some extent, the symbols and the methods of calculation. There is, however, general agreement that the portion of the serum-activity curve which follows the 3rd to the 5th day and extends to about

the 10th day is of greatest physiological significance for the study of the kinetics of the bone mineral, that this portion of the curve may be represented by a straight line on a semilog plot, and that, at least under certain conditions, it may be extrapolated to $t = 0$. Bauer *et al.*[3] have published detailed instructions for obtaining data and for making calculations using 5-day and 10-day values of (1) the specific activity of the blood, (2) the integrated blood specific activity, and (3) the retention of the isotope in the body. Other workers have published variations of these procedures.

TABLE 14.1
ACCRETION RATES UNDER NORMAL AND PATHOLOGICAL CONDITIONS[a]

| Condition | Number of observations | Accretion rate ($A$) in grams calcium per day | S.E. of mean |
|---|---|---|---|
| Normal | 24 | 0.47 | 0.03 |
| Osteopenia | | | |
| Idiopathic | 32 | 0.33 | 0.02 |
| Rheumatic | 12 | 0.37 | 0.04 |
| Post-gastric resection | 12 | 0.53 | |
| Osteogenesis imperfecta | 1 | 0.24 | |
| Steatorrhea with osteomalacia under vitamin D therapy | 1 | 2.34 | |
| Renal tubular defects | 1 | 0.41 | |
| under vitamin D therapy | 1 | 0.67 | |
| Hepatic cirrhosis | 9 | 0.34 | — |
| Endocrinopathies | | | |
| Parathyroid adenoma with osteitis fibrosa generalisata | 1 | 2.00 | |
| Acromegaly | 1 | 0.54 | |
| Cushing's disease | 1 | 0.18 | |
| Thyrotoxicosis | 1 | 0.82 | |
| | 1 | 0.83 | |
| Myxedema, untreated | 1 | 0.15 | |
| Hypogonadism (males) with | 1 | 0.57 | |
| osteopenia | 1 | 0.67 | |
| Paget's disease | 1 | 1.75 | |
| Osteosclerosis (twins) | 1 | 0.60 | |
| | 1 | 0.61 | |
| with myelofibrotic anemia | 1 | 2.07 | |
| Bone tumors | | | |
| Myeloma | 1 | 0.39 | |
| | 1 | 0.49 | |
| Metastatic carcinoma (prostate) | 1 | 6.34 | |

[a] From data of Dymling, *Medical Uses of* $Ca^{47}$.[4]

## 3. Kinetic Studies — Results

From the data in the literature, the accretion rate ($A$) of calcium in normal adult man has been calculated as approximately 0.5 g per day and the total exchangeable calcium ($E$)[3] as about 5 g. A recent report is that of Dymling,[4] in which accretion, excretory clearance rates, and exchangeable spaces are given as measured with $Ca^{47}$ and $Sr^{85}$ under normal and pathological conditions. A tabulation of the results obtained for accretion rates is given in Table 14.1. Findings in patients with bone disease are summarized in Table 14.2.

TABLE 14.2
CALCIUM ACCRETION IN BONE DISEASE[a]

| |
|---|
| *High* |
| Paget's disease |
| Fracture |
| Tumor |
| Hyperparathyroidism |
| Hyperthyroidism |
| Vitamin D-resistant rickets treated with massive doses of vitamin D |
| *Normal* |
| Osteopenia (osteoporosis) of unknown origin |
| Vitamin D-resistant rickets |
| Vitamin D-deficient rickets after treatment with vitamin D |
| *Low* |
| Osteopenia of unknown origin |
| Hypoparathyroidism |
| Hypothyroidism |
| Vitamin D-deficient rickets (untreated) |

[a] From Bauer *et al.*, *Mineral Metabolism*.[5]

## 4. Localization by External Counting

Bauer and Wendeberg[6] reported on the use of gamma-emitting isotopes of calcium ($Ca^{47}$, half-life 4.9 days) and of strontium ($Sr^{85}$, half-life 65 days) for tracing the injected isotope by means of scintillation detectors located on the body surface and predicted that the use of these techniques would assume increased importance for clarification of the kinetics of calcium metabolism under normal and pathological conditions. This prediction is in process of ample fulfillment. In addition to the contributions noted in the preceding

section, localization by external counting is coming into more general use. An increased uptake of $Ca^{47}$ and $Sr^{85}$ has been detected by external counting over metastases in the spine, over fractures, over localized lesions of Paget's disease, and over the lesions of osteitis and of osteomyelitis.[7]

External-counting studies of $Ca^{47}$ and $Sr^{85}$ have been reported from some 600 patients.[4,8] In the *Malmö allmänna sjukhus* (Malmö, Sweden), the scanning procedure has now been made routine and is conducted by the radiological laboratory, especially in the search for cancer metastases to the vertebral column in patients with negative X-ray findings. Waldeskog *et al.*,[9] reporting from that laboratory, state that 250 examinations have been made of 90 patients with breast carcinoma. Among these, 18 were positive to scanning of the vertebral column but negative to X-ray. The method is also used to record changes in isotope uptake following therapeutic irradiation or hormone therapy.

## 5. Whole-Body Counting

We have reviewed (Chapter 7) the application of whole-body counting to kinetic analysis of $Sr^{85}$ metabolism. For the purposes of this discussion, the most significant contribution is the report[8] of the possibility of detecting the retention of gamma-emitting isotopes by the whole body at desired intervals without the necessity of repeated counts of the urine and feces. (See Table 7.1).

## 6. Clinical Interpretations

From the above sections it will be seen that the figures found for values of the accretion rate ($A$) or bone formation rate ($BFR$) have proved of most value in clinical studies of the over-all skeletal metabolism and that localization of lesions by means of external counting has also proved useful, either when used alone or in combination with metabolic studies. The value for the exchangeable mineral ($E$), or the miscible calcium pool, tends to vary directly with the values for $A$, and has of itself limited clinical usefulness.

Significant deviations from the normal values for accretion have been listed in Table 14.2. Of these, the following warrant special attention from the clinical point of view:

Paget's disease
Fractures
Tumor, especially metastases to skeleton
Osteoporosis (osteopenia)
Vitamin D-deficient rickets
Vitamin D-resistant rickets

### 6.1. Paget's Disease (Osteitis Deformans)

The data obtained from studies of Paget's disease by various workers have been tabulated[5]—these data are reproduced in Table 14.3. In this condition the bone tissue is being rebuilt very rapidly, as shown by microscopic examina-

TABLE 14.3
ACCRETION AND EXCHANGEABLE MINERAL IN PAGET'S DISEASE[a]

| Patient | Investigator | Age (yrs) | Weight (kg) | Sex | *A* (g Ca/day) | *E* (g Ca) |
|---|---|---|---|---|---|---|
| HB | Spencer *et al.* (1957) | 63 | — | M | 1.4 | 9 |
| AW | Krane *et al.* (1956) | 67 | 44 | F | 3.4 | 10 |
| Mr C | Milhaud and Aubert (1958) | 55 | — | — | 5.2 | 31 |
| CG | Heaney and Whedon (1958) | 63 | 76 | M | 7.8 | 33 |
| OH | Krane *et al.* (1956) | 59 | 54 | F | 8.4 | 18 |
| EC | Heaney and Whedon (1958) | 56 | 57 | F | 8.8 | 23 |
| Normal | Bauer *et al.* (1958) | — | — | — | 0.5 | 5 |

[a] From Bauer *et al.*, *Mineral Metabolism*.[5]

tion,[11] and the serum level of alkaline phosphatase is also increased. The data in Table 14.3 are of value since they confirm the increased turnover of bone, as observed by other methods, and since they illustrate the reliability of data obtained by the kinetic methods herein described.

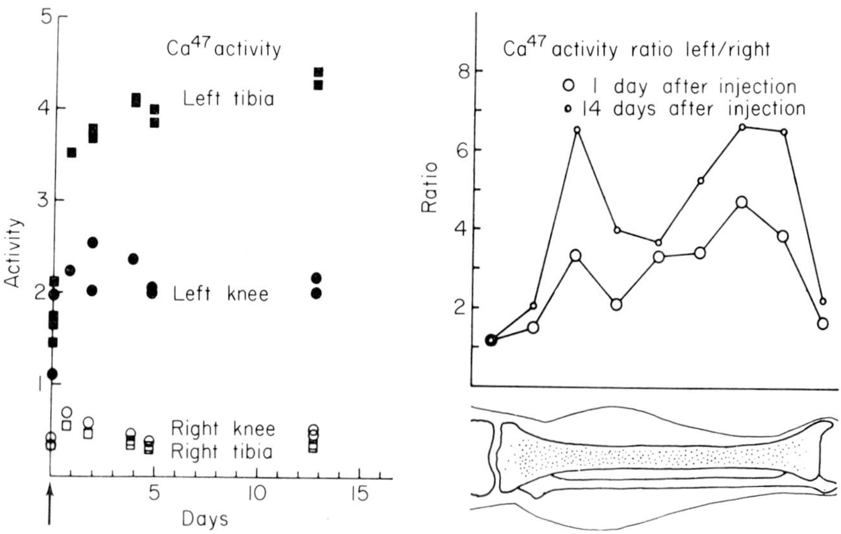


FIG. 14.1. External counting following intravenous administration of $Ca^{47}$ to a 54-year-old woman with a Paget's disease lesion in her left tibia. In agreement with the findings in studies of the accretion rate, the localized lesion showed a markedly high isotope uptake. (From Bauer and Wendeberg, *J. Bone Joint Surg.*[6])

Although it was not specifically stated, it is assumed that the patients referred to in Table 14.3 had a generalized form of Paget's disease; in any event the data represent the over-all skeletal metabolism. Data have been reported from external counting over the left tibia of a woman with Paget's disease localized in this bone.[6] The results are illustrated in Fig. 14.1. In agreement with the data for the over-all accretion rate in generalized Paget's disease, these observations illustrate a markedly high isotope uptake by the localized lesion.

### 6.2. Fractures

By the use of external counting Wendeberg[12] has demonstrated in 51 human subjects that following fracture of the tibial shaft the rate of bone salt formation in the fracture area is rapidly increased to a peak value of about 20 times normal and then gradually falls. The peak value is reached at about one-half to one year following the fracture.

### 6.3. Tumor Metastases to the Skeleton

External-counting techniques have permitted a diagnosis of metastatic cancer in the spine before X-ray evidence is positive.[7] Corey *et al.*[13] have applied kinetic analysis and balance studies to malignancy by the use of $Ca^{47}$ and $Sr^{85}$, external counting, and assays of blood and excretions for a period of 6 to 9 days. All of their calculations were based on the part of the blood-disappearance curve between the 2nd and the 6th day, extrapolating to $t = 0$. In three patients with no evidence of bone disease, the accretion rate varied from 0.34 to 0.67 g per day and the exchangeable calcium from 2.6 to 6.1 g. These values are in good agreement with those of Bauer *et al.*[3,5] In all patients with progressive metastatic disease and in one with Paget's disease, the accretion rate was elevated; the external-counting rate indicated elevated uptake over the lesions. The localized elevation of uptake was regarded as sufficient to explain the alterations in over-all kinetics of calcium, without postulating a systemic effect of the tumor on the entire skeleton. In patients with slowly progressive metastatic disease, the accretion rate was normal, and elevated uptakes were demonstrable over only a few of the lesions, which was taken to indicate that the abnormal uptake of $Ca^{47}$ was not large enough to influence the over-all accretion rate significantly. It was concluded that $Ca^{47}$ results yield information about aspects of calcium metabolism not revealed by the calcium-balance technique. Local $Ca^{47}$ uptakes reflect local calcium metabolism; the accretion rate measures the whole skeletal uptake, and the calcium balance measures the net calcium turnover of the entire body.

Corey *et al.*[14] have reported on the application of a high-energy gamma scanner (HEG scanner) to the detection of bone metastases in patients with breast cancer where the usual roentgenographs had not revealed the bone

lesions. In order to realize the unique advantages of high-energy gamma-emitting isotopes, including $Ca^{47}$ and $Sr^{85}$, a heavily shielded scanner was designed and built. The instrument was equipped with a variety of collimators (including a tungsten-focusing collimator) and with a quantitative read-out system. The entire scanning system was designed to be semiautomatic in operation and recording.

The authors report on six patients who had been given intravenous injections of either $Ca^{47}$ or $Sr^{85}$. In four of these subjects there was no prior evidence of bone metastases. Two of the patients had pain in the thoracic spine. Elevated uptakes were detected over the painful areas, where no lesions were previously detectable by X-rays. The presence of metastases was confirmed in one instance by surgery and in the other after 5 months by X-ray examination. The scans of these and other lesions demonstrate that the HEG scanning system, with its quantitative read-out, has advanced the possibility of detecting and outlining bone lesions and has also made it possible to obtain more precise data on the calcium turnover in the lesions.

Kofman *et al.*[15] have reported on the use of $Sr^{85}$ in the evaluation of bone metastases, in studies the objective of which was the development of a simple and practical method that would be useful in early diagnosis and evaluation of therapy in patients with bone metastases. These workers administered 20–60 μc of $Sr^{85}$ intravenously and then scanned the spinal column for vertebral metastases. In a study of 24 subjects with six repeat observations, they found that patients with mixed or osteoblastic metastases had a higher uptake of $Sr^{85}$ than was found in normal subjects, while in osteolytic metastases the uptake was either normal or increased. Changes in uptake of $Sr^{85}$ as a result of therapy with radiation or hormones could be demonstrated within 1 to 2 weeks. Kofman and co-workers point out that there are now several methods for studying patients with bone metastases. Each method is applicable to a different aspect of bone metabolism and each has its own advantages as well as limitations. They list the following methods: roentgenograms; calcium-balance studies; biopsies, in combination with histological examination and autoradiographs; kinetic studies, combined with calculation of accretion rates; and external scanning.

### 6.4. Osteoporosis (Osteopenia)

Perhaps more effort has been put into kinetic studies of osteoporosis than of any other condition, but the results so far reported are equivocal. Values for $A$ have been found to be high, normal, or low. The only consistent results, indicating a pronounced deviation from normal rates of bone formation and resorption, have been reported by Heaney,[16] who made observations on patients with disuse osteoporosis of poliomyelitis. These patients exhibited increased rates of bone formation and even greater increases of bone resorption. Heaney

interprets this to mean that removal of mechanical stress and strain is not associated with decreased bone-forming activity as had commonly been assumed. Although his views are contrary to accepted concepts, Heaney states that this observation on quadriplegic poliomyelitis patients is in accord with observations by Bauer[17] and by Bauer and Carlsson[18] on the osteoporosis following fracture.

From reports in the literature, it has been concluded[5] that the *A* values in osteopenia of unknown origin may be either normal or low (Table 14.2). Of six patients studied by Heaney and Whedon,[2] all had bone-formation rates within the normal range; the results obtained by Van Dilla and Arnold[19] were similar. Patients with comparable conditions in which the *A* value was lower than normal were reported by others.[20,21]

We have referred in Chapter 7 to the work of Nordin and his collaborators which, in principle, follows the contributions of Heaney and Whedon.[2] By using $Ca^{47}$ Nordin *et al.*[21] have studied seven normal subjects and seven osteoporotic patients and conclude that the calculations of Heaney and Whedon systematically overestimate the rate of new bone formation, owing to a slow diffusion component in tissue fluid-specific activities. They extrapolated the urine specific-activity curves between day 2 and day 7 to $t = 0$ and calculated the bone formation rate and the miscible calcium pool. The accretion rates ranged from 221 to 772 mg per day in the normal subjects and were not appreciably different in the patients with osteoporosis. The exchangeable pool ranged from 48 to 116 mg per kilogram in the normal subjects, and the values in osteoporosis were of the same order. Subsequent studies[22] by Nordin and MacGregor, using a variety of refinements of the methods for calculation, likewise failed to disclose significant differences between three groups of subjects—normal, osteoporotic, and steroid treated.

MacDonald[23] has introduced the radioisotope osteogram, a term applied to external counting of gamma radiation over selected areas of the skeleton, following intravenous administration of $Sr^{85}$. Special attention is given to the curve of radioactivity during the first hour. The osteograms of five patients with osteoporosis exhibited a delayed accumulation of $Sr^{85}$ over cortical bone. In these patients it required 35 minutes to reach maximum activity over the mid-tibia region as compared to 25 minutes in six normal subjects.

MacDonald[10] has also studied nine osteoporotic patients by whole-body counting. Of these, five lost 50% of the initial dose of $Sr^{85}$ within 1 to 4 days after injection of a single dose, while, in the other four subjects, 5 to 7 days were required to reach the same end point. MacDonald speculated that there may be a difference in the state of activity of the osteoporotic condition; it may be active in those patients that retain $Sr^{85}$ poorly but in a state of arrest in those with more prolonged retention. In the same study, retention of $Sr^{85}$ was plotted for as long as 100 days or more in three patients.

Urist *et al.*[24] have explored various approaches to the problem of osteoporosis, incorporating within their study the data obtained by MacDonald by the use of the radioisotope osteogram.[23] In addition, Urist and co-workers made use of metabolic-balance studies and the tetracycline labeling of bone, followed by cortical bone biopsy. They conclude that the experimental and clinical literature of the past 10 years as well as their own studies on patients indicate that prolonged calcium deficiency, castration, hyperadrenal corticoidism, or sedentary life may precipitate, accentuate, and accelerate osteoporosis in individuals who are genetically predisposed to develop it. In some instances, high calcium intake or sex hormones or both may have slowed the rate of resorption but did not replace the deficit in cortical bone.

It is difficult to arrive at any generalizations from tracer studies on osteoporosis. Heaney[25] has, however, summarized his clinical studies with radiocalcium, as follows.

1. Disuse osteoporosis in poliomyelitis, as noted above, is associated with an increased rate of bone formation and a greater increase of bone resorption.

2. The pool specific-activity curves in postmenopausal osteoporosis are not distinguishable from those of normal subjects, and calculated bone formation rates are correspondingly normal.

3. Kinetic and balance studies support a suggestion that negative calcium balance may be the cause rather than the result of postmenopausal osteoporosis.

4. Studies on estrogens and androgen analogs indicate that the positive balance effect produced by these agents is not anabolic but is anticatabolic (antiresorptive) instead.

Urist[26] suggests that osteoporosis reflects a disorder in the metabolic processes concerned with turnover of substances stored in bone tissues and believes, as opposed to Heaney, that negative calcium balance is the result rather than the cause of osteoporosis. One might equally well conclude that osteoporosis represents a shrinkage of the stable portion of bone (as represented by stable bone mineral), while the metabolic fraction of the mineral continues to function within normal limits. This view is supported by the relatively normal values for bone accretion and by other aspects of calcium metabolism such as the normal levels of serum calcium, phosphorus, alkaline phosphatase, and 24-hour urinary excretion.[27] We are unable, however, to suggest a mechanism which causes the stable portion of bone to shrink while the metabolism of bone, as represented by the labile or metabolic fraction, continues to function in a normal fashion. It can only be said that kinetic studies of skeletal metabolism yield results consistent with this hypothesis.

It is a source of some disappointment that the use of radioisotopes in the

clinical investigation of osteoporosis has not resulted in the emergence of a clearer picture of the pathological physiology of this disorder. It is important, however, to have the information that the turnover of the bone mineral in osteoporosis does not differ markedly from that in normal individuals, since future progress in the understanding of this condition must take this into account. Albright and Reifenstein[11] were explicit in their designation of osteoporosis as a deficiency in bone matrix protoplasm, and nothing in the reports cited above contradicts this view. It should not have been expected that tracer studies with bone-seeking radioisotopes would elucidate the failure of the skeleton to respond to such a deficiency with a vigorous repair reaction.

## 6.5. Vitamin D-Deficient Rickets

Vitamin D-deficient rickets has been studied in infants by the use of $P^{32}$ as a tracer.[28] The accretion rate of bone salt was found to be lower than in normal children of the same age; administration of vitamin D restored the rate to normal. It was concluded that treatment of rachitic children with vitamin D causes not only an increased accretion rate but also increased resorption. The same conclusions have been reached by the same authors in the study of rachitic rats, and their results have been reviewed in Chapter 9.

Migicovsky[29] and Migicovsky and Bélanger,[30] who studied rachitic chicks by the use of $Ca^{45}$ as a tracer element, have arrived at different conclusions. They agree that vitamin D increases the absorption of calcium from the intestinal tract and have demonstrated that this effect takes place a short time after the vitamin is administered. These investigators also conclude that there is a direct effect upon bone, but they believe this effect to be on accretion and not on resorption. Gran,[31] by comparing the data obtained from rats deprived of vitamin D with those given the vitamin, also concluded that vitamin D acts primarily by increasing absorption and accretion of calcium. The nature of the peripheral effect of vitamin D, if any, on the healing of vitamin D-deficient rickets thus remains in dispute.

## 6.6. Vitamin D-Resistant Rickets

This condition is, by definition, one which does not heal when treated with the usual therapeutic doses of vitamin D, and this failure is ordinarily accepted as diagnostic. Fanconi,[32] however, describes four forms of vitamin D-resistant rickets. According to his criteria, primary vitamin D-resistant rickets (which he also calls phosphate diabetes) is caused by insufficient reabsorption of phosphate in the proximal renal tubules and results in a low blood phosphate. The condition responds poorly or not at all to vitamin D, even in large doses; the $Ca \times P$ product in the serum remains low, and the rachitic condition in the bones persists.

Bauer *et al.*[28] studied two children with primary vitamin D-resistant rickets

by using $P^{32}$ as a tracer element. They found the accretion rate within normal limits, but after treatment with massive doses of vitamin D this rate reached values higher than those in normal children of the same age.

Two patients with deformities of the long bones, diagnosed as vitamin D-resistant rickets on the basis of characteristic roentgenographic changes, and one normal control have been studied with $Sr^{85}$ by using kinetic analysis.[33] Plasma clearance of the isotope by excretion and by accretion was calculated. The two patients with vitamin D-resistant rickets exhibited a rapid rate of clearance by accretion.

## References


1. G. C. H. Bauer, A. Carlsson, and B. Lindquist, Evaluation of Accretion Resorption and Exchange Reactions in the Skeleton, *Kungl. Fysiol. Sällskapets Lund Förhandl.*, **25**: 1-16 (1955).
2. R. P. Heaney and G. D. Whedon, Radiocalcium Studies of Bone Formation Rate in Human Metabolic Bone Disease. *J. Clin. Endocrinol. Metabol.*, **18**: 1246-1267 (1958).
3. G. C. H. Bauer, A. Carlsson, and B. Lindquist, Use of Isotopes in Clinical Studies of Skeletal Metabolism, in *Radioaktive Isotope in Klinik und Forschung*, III, pp. 25-39, Urban & Schwarzenberg, München, 1958.
4. J. F. Dymling, Accretion and Excretory Clearance Rates and Exchangeable Spaces Measured in Man with $Ca^{47}$ and $Sr^{85}$ Under Normal and Pathological Conditions, in *Medical Uses of $Ca^{47}$*, IAEA Technical Reports Series No. 10, pp. 73-76, Intern. Atomic Energy Agency, Vienna, 1962.
5. G. C. H. Bauer, A. Carlsson, and B. Lindquist, Metabolism and Homeostatic Function of Bone, in *Mineral Metabolism*, Vol. I, pp. 609-676, Academic Press Inc., New York, 1961.
6. G. C. H. Bauer and B. Wendeberg, External Counting of $Ca^{47}$ and $Sr^{85}$ in Studies of Localised Skeletal Lesions in Man, *J. Bone Joint Surg.*, **41-B**: 558-580 (1959).
7. G. C. H. Bauer, Kinetics of Calcium and Strontium Metabolism in Man, in *Bone as a Tissue*, pp. 118-127, McGraw-Hill Book Co., Inc., New York, 1960.
8. G. C. H. Bauer, Experience with External Counting Techniques in Clinical Studies of Bone, in *Medical Uses of $Ca^{47}$*, IAEA Technical Reports Series No. 10, pp. 92-95, Intern. Atomic Energy Agency, Vienna, 1962.
9. B. Waldeskog, I. Gynning, and P. Langland, Localization with $Sr^{85}$ of Spinal Metastases in Mammary Cancer and Changes in Uptake after Hormone and Roentgen Therapy, in *Medical Uses of $Ca^{47}$*, IAEA Technical Reports Series No. 10, pp. 104-106, Intern. Atomic Energy Agency, Vienna, 1962.
10. N. S. MacDonald, Recent Uses of a Total-body Counter Facility in Metabolic Research and Clinical Diagnosis with Radionuclides, in *Whole-body Counting*, pp. 502-514. Intern. Atomic Energy Agency, Vienna, 1962.
11. F. Albright and E. C. Reifenstein, Jr., *Parathyroid Glands and Metabolic Bone Disease*, Williams & Wilkins Co., Baltimore, 1948.
12. B. Wendeberg, Mineral Metabolism of Fractures of the Tibia in Man Studied with External Counting of $Sr^{85}$, *Acta Orthopaed. Scand., Suppl. 52* (1961).
13. K. R. Corey, P. Kenny, E. Greenberg, and J. S. Laughlin, Results of Kinetic Analyses and Calcium Balance Studies in Malignancy, in *Medical Uses of $Ca^{47}$*, IAEA Technical Reports Series No. 10, pp. 80-84, Intern. Atomic Energy Agency, Vienna, 1962.

14. K. R. Corey, P. Kenny, E. Greenberg, and J. S. Laughlin, Detection of Bone Metastases in Scanning Studies with Calcium-47 and Strontium-85, *J. Nucl. Med.* **3**: 454-471 (1962).
15. S. Kofman, H. H. Sky-Peck, Y. Thibaudeau, R. D. Ray, and S. G. Taylor, III, The Use of Strontium-85 in the Evaluation of Bone Metastases: A Preliminary Report, *J. Nucl. Med.* **4**: 9-17 (1963).
16. R. P. Heaney, Summary of Results from Clinical Studies with Radiocalcium, in *Medical Uses of $Ca^{47}$*, IAEA Technical Reports Series No. 10, pp. 77-84, Intern. Atomic Energy Agency, Vienna, 1962.
17. G. C. H. Bauer, Rate of Bone Salt Formation in a Healing Fracture Determined in Rats by Means of Radiocalcium, *Acta Orthopaed. Scand.*, **23**: 169-191 (1954).
18. G. C. H. Bauer and A. Carlsson, Post-fracture Bone Salt Resorption Studied in Rats, *Acta Orthopaed. Scand.*, **25**: 83-88 (1955).
19. M. A. Van Dilla and J. S. Arnold, $Strontium^{85}$ Tracer Studies in Humans, *Intern. J. Appl. Radiation Isotopes*, **1**: 129 only (1956).
20. G. Milhaud and J. P. Aubert, Étude des Voies Générales du Métabolisme Calcique chez l'Homme, in *Radioisotopes in Scientific Research*, Vol. IV, pp. 120-124, Pergamon Press, New York, 1958.
21. B. E. C. Nordin, M. Bluhm, and J. MacGregor, *In vitro* and *in vivo* Studies with Bone-seeking Isotopes, in *Radioisotopes and Bone*, pp. 105-126, Blackwell Scientific Publ., Oxford, 1962.
22. B. E. C. Nordin and J. MacGregor, Observations on Calcium Kinetics in Osteoporosis and in Steroid-treated Subjects, in *Medical Uses of $Ca^{47}$*, IAEA Technical Reports Series No. 10, pp. 78-80, Intern. Atomic Energy Agency, Vienna, 1962.
23. N. S. MacDonald, The Radioisotope Osteogram—Kinetic Studies of Skeletal Disorders in Humans, *Clin. Orthop.*, **17**: 154-166 (1960).
24. M. R. Urist, P. S. Zaccalini, N. S. MacDonald, and W. A. Skoog, New Approaches to the Problem of Osteoporosis, *J. Bone Joint Surg.*, **44-B**: 464-484 (1962).
25. R. P. Heaney, Evaluation of Calcium Kinetics in Adult Humans, in *Medical Uses of $Ca^{47}$*, IAEA Technical Reports Series No. 10, pp. 48-51, Intern. Atomic Energy Agency, Vienna, 1962.
26. M. R. Urist, Osteoporosis, *Ann. Rev. Med.*, **13**: 273-286 (1962).
27. N. F. Moon and M. R. Urist, Gerontal and Gerontoid Osteoporosis: A Classification Based on a Survey of Cases Investigated Recently in a University Hospital, *Clin. Orthopaed.*, **23**: 269-282 (1962).
28. G. C. H. Bauer, A. Carlsson, and B. Lindquist, Bone Salt Metabolism in Human Rickets Studied with Radioactive Phosphorus, *Metabolism*, **5**: 573-581 (1956).
29. B. B. Migicovsky, Influence of Vitamin D on Calcium Resorption and Accretion, *Can. J. Biochem. Physiol.*, **35**: 1267-1275 (1957).
30. B. B. Migicovsky, and L. F. Bélanger, Rickets Studied with Isotopes, in *Radioisotopes and Bone*, pp. 167-189, Blackwell Scientific Publ., Oxford, 1962.
31. F. C. Gran, Studies on Calcium and Strontium-90 Metabolism in Rats, *Acta Physiol. Scand.*, **48**, *Suppl. 167* (1960).
32. G. Fanconi, Variations in Sensitivity to Vitamin D: from Vitamin D Resistant Rickets, Vitamin D Avitaminotic Rickets and Hypervitaminosis D to Idiopathic Hypercalcaemia, in *Ciba Foundation Symposium on Bone Structure and Metabolism*, pp. 187-205, Little, Brown & Co., Boston, 1956.
33. W. Meltzer, I. Lyon, E. D. Mensen, and R. D. Ray, Radioisotope Studies of Generalized Skeletal Disorders: Vitamin D Resistant Rickets, *Clin. Orthopaed.*, **17**: 269-287 (1960).

# AUTHOR INDEX

Numbers in parentheses are included to assist in locating a reference when the author's name is not cited in the text. Numbers in italics indicate the page on which the complete reference is listed.

## A

Albarede, P., 140 (21), *145*
Albright, F., 191 (11), 196, *197*
Amprino, R., 132, *144*
Archibald, R. M., 151 (10), 153 (10), *157*
Armstrong, W. D., 66, 67, *76*, 85 (19), *89*, 91, *102*, 124, 125, *129*
Arnold, J. S., 44 (18), 45 (18), *46*, 88 (33), *90*, 148, 149, *157*, 194, *198*
Asling, C. W., 33 (7), *45*, 81, 85 (14), 88 (14), *89*
Atherton, D. R., 101 (32), *104*
Aub, J. C., 131, *144*, 167, *169*
Aubert, J. P., 97 (22), *103*, 194 (20), *198*

## B

Bacon, J. A., 126, *129*
Baden, H., 80 (3), *88*
Bauer, G. C. H., 80 (4), *89*, 92, 93 (8), 94, 95, 96, 99, 101, *102*, 105 (1), 114, *116*, 127, *130*, 140 (28), *145*, 168 (13), *169*, 187, 188, 189, 190 (78), 191 (5), 192 (6, 7), 194 (5), 196 (28), *197*, *198*
Bear, R. S., 15 (12), 17, *29*
Behrens, C. F., 48 (8), *60*
Bélanger, L. F., 27 (25), *30*, 31 (6), *45*, 153 (10), 155 (15), *157*, 196, *198*
Belchier, J., 131, *144*
Benda, C. E., 92 (7), *102*
Bender, M. A., 140 (30), *145*
Bergstrand, P. J., 169 (15), *169*
Bird, E. D., 82 (17), *89*
Biskis, B. O., 169 (15), *169*
Björnerstedt, R., 132 (13), 134 (13), *144*
Blau, M., 140, *145*
Bloom, M. A., 14 (7), 18 (15), *29*
Bloom, W., 14 (6, 7), 18 (15), 19, 20, 22, 26 (19), 27 (19), *29*, *30*, 60 (13), *60*, 122 (2), *128*
Bluhm, M., 97 (19), *103*, 194 (21), *198*
Blum, T., 1, *11*
Bobb, F. C., 33 (8), *46*
Booth, C. C., 81 (9), *89*
Borle, A. B., 74, *77*
Boyd, G. A., 132, *144*
Bradford, R. H., 125 (16), *129*
Brasseur, H., 64 (9), *76*
Breibart, S., 81, *89*
Bronner, F., 92, *102*, 126, *129*, 151 (10), 153 (10), *157*
Brothers, M., 114 (20), *117*
Brues, A. M., 100 (29), *103*
Buchanan, D. L., 39 (13), *46*, 79 (1), *88*
Budy, A. M., 14 (8), *29*, 157, *158*
Burger, M., 72 (32), *77*

## C

Cameron, E. C., 122 (5, 6), *128*
Campo, R. D., 153 (12), 156, *157*, *158*
Carlson, C. H., 85 (19), *89*
Carlsson, A., 92 (8, 9), 93 (8), 94 (8), 95 (8, 10), 101, *102*, 105 (1), 113, 114 (1), *116*, *117*, 127 (29), *129*, *130*, 168 (13), *169*, 187 (1), 188 (3), 189 (5), 191 (5), 192 (3, 5), 194 (5), 196 (28), *197*, *198*
Carneiro, J., 156 (23), *158*
Carr, T. E. F., 182, *185*
Causeret, J., 128, *130*
Chase, G. D., 48 (3), *60*
Chen, P. S., Jr., 31 (1), 33 (1), *45*, 85 (23), *89*, 124 (11), *129*
Cheney, B. A., 122 (7), *128*
Chiewitz, O., 1, *11*, 91, *102*, 131, *144*
Childs, D. S., Jr., 167 (11), *169*
Clark, H. M., 48 (1), 49 (1), 50 (1), 51 (1), 54, *60*, 171 (3), *184*
Clark, I., 126, *129*
Clemedson, C. J., 132 (13), 134 (13), *144*
Cohen, J., 99, *103*
Cohn, S. H., 178 (6), *184*
Comar, C. L., 48 (2), *60*, 85 (21), *89*, 91 (6), 92 (6), *102*, 109, 110 (6), 112 (13), 114, 115, 116 (21), *116*, *117*, 132, 136 (11), 137, 138, *144*

Conard, R. A., 178 (6), *184*
Copp, D. H., 123 (5), 126 (21), *128, 129*
Corey, K. R., 140, *145*, 192, *197, 198*
Cotzias, G. C., 86, *89*
Craig, L. C., 122, *128*
Cramer, C. F., 126, *129*

## D

Dahl, L. K., 81 (8), *89*
Dahlin, D. C., 167 (11), *169*
Dallemagne, M. L., 64 (9, 10), 66 (18), 67, 68, *76*
Darby, W. J., 82 (18), *89*
Davidson, A. G. F., 122 (6, 7), *128*
Davis, G. K., 91 (6), 92 (6), *102*
De Morgan, C., 21, *29*
Dickens, F., 69, *76*
Di Ferrante, N., 151 (10), 153 (10), *157*
Di Stefano, V., 124 (11), *129*
Dobyns, B. M., 136, *145*
Dorfman, A., 13 (1), 16 (1), *29*
Dowdle, E. B., 111 (8), *117*
Dowse, C. M., 74 (35), *77*, 124 (10, 13), *128, 129*
Duckworth, J., 31 (2), 33 (2), *45*, 81, *89*
Dudley, H. C., 140, *145*
Dudley, R. A., 136, *145*, 159 (4), *169*
Duhamel, H. L., 131, *144*
Dunham, C. L., 167 (8), *169*
Dunning, G. M., 180, 183 (9), *185*
Durbin, P. W., 31 (4), 33 (4), *45*, 39 (12), *46*, 85, 86 (25), *89*, 151 (6), *157*
Duthie, R. B., 153 (14), *157*
Dymling, J. F., 97, *103*, 188, 190 (4), *197*
Dziewiatkowski, D. D., 151, 153 (10, 12, 13), 156, *157, 158*

## E

Edelman, I. S., 80 (3), *88*
Eeg-Larsen, N., 113 (17), *117*, 127, 128, *129, 130*
Egnund, K. M., 113 (18), *117*
Elliott, S. R., 16 (13), *29*
Engström, A., 62 (2), 63 (7), 68, *75, 76*, 132, 134, *144*
Ericsson, Y., 84, *89*
Evans, R. D., 51 (12), *60*, 159 (3), 167 (10), *169*

## F

Faber, M., 45, *46*
Fabry, C., 64 (10), 65 (14), 66 (14), *76*
Fairhall, L. T., 131 (4), *144*
Farid, Z., 82 (18), *89*
Fawcett, D. W., 22, 26 (19), 27 (19), *30*
Feitelberg, S., 48 (5), *60*
Finean, J. B., 63 (7), *76*
Finkel, M. P., 166 (7), 169 (15), *169*
Firschein, H., 124 (11), *129*
Floyd, R. L., 101 (32), *104*
Foland, W. D., 33 (8), *46*
Forbes, G. B., 80 (6), 81 (10), *89*
Fore, H., 86 (26), *89*
Francis, G. E., 48 (4), *60*, 138, *145*
François, P., 65 (14), 66 (14, 16, 17), *76*
Franconi, G., 196, *198*
Frank, R. M., 70 (26), *76*
Fuerst, C. R., 37 (11), *46*

## G

Gaillard, P. J., 122 (3), *128*
Gee, M., 39 (12), *46*, 151 (16), *157*
Gersh, I., 13 (3), *29*
Geschickter, C. F., 91 (4, 5), 99 (4), *102*
Gibbs, W. D., 81, *89*
Glaser, W., 81, *89*
Glimcher, M. J., 62, 72 (29, 30, 31), *75, 77*
Godden, W., 81 (13), *89*
Gran, F. C., 196, *198*
Greenberg, D. M., 127, *129*, 131, *144*
Greenberg, E., 140 (27), *145*, 192 (13, 14), *197, 198*
Greulich, R. C., 27 (25), *30*, 31 (6), *45*, 153 (10), *157*
Gross, J., 15 (11), 16, *29*, 156 (24), *158*
Guggenheim, E. A., 7 (5), *11*
Gynning, I., 190 (9), *197*

## H

Haavaldsen, R., 113, *117*
Hamada, G. H., 116 (21), *117*
Hamilton, J. G., 39 (12), *46*, 151 (6), *157*
Hanna, S., 81 (9), *89*
Hansard, S. L., 91, 92, *102*, 126 (22), *129*
Harris, R. S., 92 (7), *102*
Harrison, G. E., 112, *117*, 182 (12), *185*
Hastings, A. B., 106, 107 (2), 108 (4), *116*, 124, *128*

Haumont, S., 82, 83, *89*
Heaney, R. P., 97, 98, *103*, 187, 193, 194, 195, *197*, *198*
Heller, M., 122, 124, *128*, 161 (5), *169*
Heller-Steinberg, M., 13 (4), *29*
Hempelmann, L. H., 167 (10), *169*
Hendricks, S. B., 64 (8), *76*
Henze, K. G., 122 (7), *128*
Herman, H., 65 (14), 66 (14, 16, 18), 67 (19), 68, *76*
Hevesy, G., 1, *11*, 91, *102*, 131, *144*
Hill, R., 31 (2), 33 (2), *45*
Hill, W. J., 64 (8), *76*
Hodge, H. C., 31 (1), 33 (1), *45*, 85 (23), *89*
Hodgman, C. D., 6, 7 (6), *11*
Hoffman, F. L., 1, *11*
Hollander, J. M., 33 (10), 35 (10), 37 (10), 38 (10), *46*, 147 (1), *157*
Hollister, H., 39 (14), *46*
Howard, R. P., 125 (16), *129*
Huggins C. B. 124, *128*
Hugot, D., 128, *130*
Humphries, R. E., 1, *11*
Hurley, L. S., 33 (7), *45*, 81, 85 (14), 88 (14), *89*

**I**

Irving, J. T., 69 (25), *76*
Ivins, J. C., 167 (11), *169*

**J**

Jackson, S. F., 72, *77*, 156 (25, 26), *158*
James, A. H., 80 (3), *88*
Jee, W. S. S., 88 (33), *90*, 148, 149, *157*
Joel, W., 125 (16), *129*
Johnson, J. A., 91 (3), *102*
Jowsey, J., 151 (7), *157*

**K**

Kallfelz, F. A., 112 (13), *117*
Kamen, M. D., 48 (6), *60*, 138, *145*
Kember, N. F., 155 (18), *158*
Kenny, P., 140 (27), *145*, 192 (13, 14), *197*, *198*
Kern, R., 70 (26), *76*
King, E. R., 48 (7), *60*
Kisieleski, W., 91, *102*
Kodicek, E., 157, *158*
Kofman, S., 193, *198*
Kramer, B., 69 (23), *76*
Krane, S. M., 72 (29, 30), *77*
Kulp, J. L., 182 (10), 184, *185*

**L**

Lacroix, P., 18 (17), *29*
Langland, P., 190 (9), *197*
Laszlo, D., 114 (20), *117*
Laughlin, J. S., 140 (27), *145*, 192 (13, 14), *197*, *198*
Law, L. L., 167 (9), *169*
Leblond, C. P., 27 (25), *30*, 31 (6), *45*, 153 (11), 156 (23), *157*, *158*
Lee, J. S., 81 (10), *89*
Lengemann, F. W., 110 (6), 112, 114 (23), *116*, *117*
Lewis, A., 80 (6), *89*
Lienke, R. I., 91 (3), *102*
Lindquist, B., 92 (8, 9), 93 (8), 94 (8), 95 (8, 10), 101, *102*, 105 (1), 114 (1), *116*, 127 (29), *129*, *130*, 168 (13), *169*, 187 (1), 188 (3), 189 (5), 191 (5), 192 (3, 5), 194 (5), 196 (28), *197*
Lippincott, E. R., 64 (12), *76*
Litovitz, T. A., 91 (4, 5), 99 (4), *102*
Lloyd, E., 151 (8), *157*
Looney, W. B., 41 (15), 45, *46*
Lotz, W. E., 85 (21), *89*
Lough, S. A., 116, *117*
Loutit, J. F., 181, 182, *185*
Lucas, H. F., Jr., 44 (17), *46*, 172 (4), *184*
Lumsden, E., 112 (12), *117*
Lyon, I., 155 (28), 156 (28), *158*, 197 (33), *198*

**M**

McConnell, D., 68 (21), 71, *76*
McCoord, A., 81 (10), *89*
McCorkle, W. C., 33 (8), *46*
MacDonald, N. S., 98, 99 (24), *103*, 142, *145*, 194, 195 (23, 24), *197*, *198*
MacGregor, J., 97 (19, 20), *103*, 194 (21, 22), *198*
MacIntyre, I., 81, *89*
McLean, F. C., 14 (5, 6, 7, 8), 20, *29*, 37, *46*, 60 (13), *60*, 61 (1), 64 (1), 66 (1), *75*, 106, 107 (2), *116*, 119 (1), 122 (2), 123, 124 (2), 127, *128*, *129*, 178, *185*

Mac Pherson, S., 155 (21), 156 (21), *158*
Macpherson, S., 168, *169*
Maletskos, C. J., 92 (7), 99 (28), *102, 103*
Malm, O. J., 113 (17), *117*
Marinelli, L. D., 44 (16, 17), *46*, 142, 144, *145, 146*
Markowitz, H. A., 140 (24), *145*
Marshall, J. H., 87, 88 (31, 32), *90*, 99 (28), 100, 101 (31), *103, 104*, 166 (7), *169*
Martin, G. R., 125 (14), *129*
Martland, H. S., 1, *11*, 167 (10), *169*
May, A. A., 144 (37), *146*
Mehl, H. G., 140, *145*
Meltzer, W., 197 (33), *198*
Mensen, E. D., 197 (33), *198*
Meyer, K., 13 (2), 16 (2), *29*
Migicousky, B. B., 196, *198*
Milch, R. A., 131, *144*
Milhaud, G., 97, *103*, 194 (20), *198*
Miller, C. E., 142, 144 (37), *145, 146*
Minot, A. S., 131 (4), *144*
Mirale, A., Jr., 82 (18), *89*
Mitchell, T. G., 48 (7), *60*, 140 (24), *145*
Mizner, G. L., 80 (6), *89*
Molnar, Z., 62 (4), *75*
Monroe, R. A., 85 (21), *89*
Moon, N. F., 195 (27), *198*
Moore, F. D., 80 (3), *88*
Morton, R. A., 86 (26), *89*
Mulligan, W., 48 (4), *60*, 138 (20), *145*
Mulry, W. G., 140, *145*
Mulryan, B. J., 124 (11), 125 (14), *129*

**N**

Nagler, W., 140 (30), *145*
Nahum, H., 140 (21), *145*
Nakao, A., 39 (13), *46*, 79 (1), *88*
Nelson, A., 132 (13), 134 (13), *144*
Neuberger, A., 157 (27), *158*
Neuman, M. W., 63, 68, 69 (6), 74 (6), 75, *75*, 107 (3), *116*, 128, *130*
Neuman, W. F., 63, 68, 69 (6, 25), 74 (6, 35), 75, *76, 77*, 107 (3), *116*, 124, 125, 128, *128, 129, 130*
Newman, R. H., 39 (12), *46*, 151 (6), *157*
Nicolaysen, R., 113 (18), *117*, 127, 128, *129, 130*
Nichols, G., Jr., 74 (37), *77*, 80 (2), *88*, 156, *158*
Nichols, N., 74 (37), *77*, 80 (2), *88*
Nobel, S., 72 (32), *77*
Nold, M. M., 110 (6), *116*
Nordin, B. E. C., 74, *77*, 97, *103*, 194, *198*
Norris, W. P., 91, 100, *102, 103*, 136 (17), *145*

**O**

Obeblad, E., 136 (18), *145*
Overman, R. T., 48 (1), 49 (1), 50 (1), 51 (1), 54, *60*, 171 (3), *184*
Owen, M., 155 (20, 21), 156 (21), *158*, 168 (12), *169*

**P**

Patrick, H., 126 (22), *129*
Perkinson, J. D., Jr., 85, *89*
Perloff, A., 64 (11), *76*
Posner, A. S., 64 (10, 11, 12), *76*
Prasad, A. S., 82, *89*
Premer, M. L., 91 (3), *102*

**Q**

Quimby, E. H., 48 (5), *60*

**R**

Rall, D. P., 131 (6), *144*
Ramsden, E. N., 151, *157*
Randall, J. T., 72, *77*
Rasmussen, N. C., 51 (12), *60*, 112, *117*, 122, 126 (20), *128, 129*, 159 (4), *169*
Ray, R. D., 95, 96, 99, *102*, 140, *145*, 155, 156, *158*, 193 (15), 197 (33), *198*
Raymond, W. H. A., 112 (12), *117*
Read, A. E., 81 (9), *89*
Reifenstein, E. C., Jr., 191 (11), 196, *197*
Reznikoff, P., 131 (4), *144*
Rigal, W. M., 155 (22), *158*
Robertson, J. S., 81 (8), *89*, 99, *103*, 178 (6), *184*
Robinson, C. V., 51 (11), 55, 57, *60*
Robinson, R. A., 16 (13), 17 (14), 18, *29*, 62 (3), 72 (3, 33), 73, *75, 77*
Rockenmacher, M., 69 (23), *76*
Rönnholm, E., 70, *76*
Rose, J. E., 144 (37), *146*
Rosen, S. M., 111 (7), *117*
Rosenthal, M. W., 183 (13), *185*

Rowland, R. E., 22 (20), *30*, 87, 88 (31), *90*, 155 (28), 156 (28), *158*
Rubin, M. I., 91 (4), 99 (4), *102*

## S

Sabanas, A. O., 167, *169*
Sandstead, H. H., 82 (18), *89*
Sargent, T. W., 143, 144 (35), *145*
Schacter, D., 111, *117*
Schenker, H., 111 (8), *117*
Schulert, A. R., 82 (18), *89*, 182 (10), 184, *185*
Scoccianti, P., 140 (28), *145*
Seaborg, G. T., 33 (10), 35 (10), 37 (10), *46*, 147 (1), *157*
Sendroy, J., Jr., 108 (4), *116*
Service, D. H., 85 (19), *89*
Seymour, A., 177, *184*
Sheldon, H., 72 (33), 73, *77*
Shetlar, M. R., 125, *129*
Silver, L., 81, *89*
Silver, S., 48 (5), *60*
Simmons, D. J., 155 (16), *157*
Simons, D. G., 171 (2), *184*
Singer, L., 66, 67, *76*, 85 (19), *89*, 91 (3), *102*
Sissons, H. A., 151 (7), *157*
Skoog, W. A., 195 (24), *198*
Sky-Peck, H. H., 193 (15) *198*
Slack, H. G. B., 156 (27), *158*
Smith, R. H., 156 (25, 26), *158*
Sobel, A. E., 69, 72 (32), *76*, *77*
Sognnaes, R. F., 23 (21), 24, 25, 26 (23), *30*, 70 (26), *76*, 86 (29), *90*
Solomon, A. K., 99, *103*
Solomons, C. C., 69 (25), *76*
Spencer, H., 114, *117*
Spiers, F. W., 141, 143, *145*
Stack, M. V., 27 (24), *30*, 71 (28), *77*
Stent, G. S., 37 (11), *46*
Stevens, J., 155 (28), 156 (28), *158*
Stover, B. J., 44 (18), 45 (18), *46*, 101 (32), *104*, 149 (3), *157*
Strominger, D., 33 (10), 35, 37, 38, *46*, 147 (1), *157*
Stutman, J. M., 64 (12), *76*
Suiker, A. P., 126 (21), *129*
Sutton, A., 112 (12), *117*
Sutton, A., 182 (12), *185*

## T

Talmage, R. V., 125, 126 (19), *129*
Taylor, S. G., III, 193 (15), *198*
Taylor, T. G., 81, *89*
Taysum, D. H., 101 (32), *104*
Terepka, A. R., 31 (1), 33 (1), *45*, 74 (35), *77*, 85 (23), *89*, 124 (13), *129*
Thibaudeau, Y., 193 (15), *198*
Thiers, R. E., 88 (30), *90*
Thomas, R. O., 91, 99 (4), *102*
Thomas, W. C., Jr., 82 (17), *89*
Thompson, R. C., 31 (3), 33 (3), *45*
Tipton, I. H., 33 (8), *46*
Tobie, J. E., 131 (6), *144*
Toft, R. P., 125 (15), 126 (19), *129*
Tomes, J., 21, *29*
Tonna, E. A., 155 (17), *157*
Totter, J. R., 39 (14), *46*
Tsuzuki, M., 178 (7), *185*
Tubiana, M., 140 (21), *145*
Twardock, A. R., 116, *117*
Tyler, S. A., 100 (29), *103*

## U

Ullberg, S., 84, *89*, 132, *144*
Underwood, E. J., 85 (24), *89*
Urist, M. R., 14 (5, 8), *29*, 37, *46*, 61 (1), 64 (1), 66 (1), *75*, 123, *128*, 178, *185*, 195 (27), *198*

## V

Vaes, G. M., 156, *158*
Vallee, B. L., 81, 88 (30), *89*, *90*
Van Dilla, M. A., 44, 45 (18), *46*, 101, *104*, 194, *198*
Vaughan, J., 31 (5), *45*, 151 (7), *157*, 166 (6), 168 (12) *169*
Veall, N., 143, *146*
Vincent, J., 80, *89*, 131, *144*

## W

Waldeskog, B., 190, *197*
Warnock, G. M., 81 (13), *89*
Wasserman, R. H., 109, 110, 112, 114, 115, *116*, *117*
Wassermann, F., 15 (9), *29*
Wendelberg, B., 96 (13), *103*, 140, *145*, 189, 191, 192 (6), *197*
Whedon, G. D., 97, *103*, 187, 194, *197*

Whitney, I. B., 85 (21), *89*, 114 (23), *117*
Williams, J. B., 99 (28), *103*
Williams, M. H., 39 (12), *46*, 151 (6), *157*
Wimer, L. T., 125 (15), *129*
Winand, L., 64, 65 (13), *76*
Woodruff, L. A., 136 (17), *145*
Woods, K. R., 124, 125, *129*
Wormall, A., 48 (4), *60*, 138 (20), *145*

## Y

Yagoda, H., 132, *144*
Young, R. W., 155 (19), *158*

## Z

Zaccalini, P. S., 195 (24), *198*
Zelle, M. R., 39 (14), *46*

# SUBJECT INDEX


## A

Accretion, 92, 93, 114, 189, 195, 197
  in Paget's disease, 191
  periosteal, 82
Accretion rate (A), 96, 97, 100, 188, 189, 190
Actinide series, 32, 33
Actinium (Ac), 33
  $Ac^{228}$ ($Mesothorium_2$), 43, 44, 149
Adenosine triphosphate (ATP), 16
Adsorption phenomena, 75
Alizarin, 31, 131
Alkaline earths, 31, 41
Alkaline phosphatase, 28, 32, 34, 82, 84, 191, 195
Alpha decay, 8, 33, 41, 42, 43, 44
Alpha particles, 8, 33, 49, 50, 58, 166, 171
  tracks of, 133, 136, 148
  transmutation induced by, 10
Alpha radiation, 88, 160, 161
Aluminum (Al), 33
Alveolar bone, 28
Ameloblasts, 25, 26
Amelogenesis, 27
Amino acids
  in bone cells, 156
  pyrrolidine, 15
  tritiated, 155-6
Amino sugars, 16
Annihilation quanta, 51, 52
Anthracene (phosphor), 53, 54
Antimony ($Sb^{124}$), 142
Apatite crystals, 23, 24, 27, 61, 63, 64, 65, 66, 67, 84, 85, 147
  calcium-deficient, 64, 65, 66, 67, 68
  carbonate-containing, 68, 71
  equilibrium between solid and liquid, 71-2
  low-calcium, 64
  orientation, 74
  stable, 68
Apatite series, 63, 147
Atomic nucleus, 3, 7, 8, 140
Atomic (Z) number of elements, 4-6, 8, 88
Atomic structure, 2-7
Atomic weights. 4-6, 7
Atoms, 2-3
Autoradiograph, 59, 82, 83, 84, 87, 134-5
  alpha track, 148
Autoradiography, 59-60, 131, 132, 134-6, 138
  qualitative, 132, 134-6
  quantitative, 136, 138

## B

Barium (Ba), 31, 36, 39, 85, 178
  $Ba^{133}$, 9
  $Ba^{137}$, 36
  $Ba^{140}$, 35, 39, 55, 178
Beryllium (Be), 31
  $Be^7$, 9, 34, 55
Beta decay, 8, 9, 33, 36, 37, 41, 42, 43, 44
  beta particles, 8, 36, 37, 49, 50, 52, 177
  $\beta-$ (negatron), 8, 34, 35, 37, 38, 42, 43, 55, 162-5
  $\beta+$ (positron), 8, 33, 34, 35, 37, 38, 52, 162
  emission of radionuclides by, 172
  tracks of, 136, 138
Beta radiation, 88, 161
Bicarbonate, $C^{14}$-labeled, 155
Bismuth (Bi), 33, 40, 131
Blood disappearance curves, exponentials, 91-2, 97-8
Blood, specific activity of, 91
Blood plasma
  anions, 119
  cations, 119
  concentration of calcium ion in, 122
Blood vessels, 21
Body fluids, see Fluids of body
Bone formation, 18-22, 98
Bone formation rate (BFR), 97, 187, 190, 194
Bone as a tissue, 13-17
Bone cells, see Osteoblasts; Osteoclasts; Osteocytes
Bone marrow, 14, 185
Bone matrix, 147-53, see also Matrix; Organic matrix
Bone mineral (bone salt), 15, 22-3, 31, 61-75, 124

anionic constituents, 66
chemical composition, 61
chemical nature, 65
crystal structure, 61
dissolution, 125
dynamics of, 71
exchangeable, 92, 190
kinetics of, 91-102
labile, 67
nonexchangeable, 93
nucleation, 72
renewal rate, 134
solubility, 71
solubilization, 74
stable, 67, 125
transfer from bone to blood, 74
Bone-seeking elements, 31-45, 91, 135
Bone surfaces, 53, 147, 184
endosteal, 149, 151, 166, 168
periosteal, 149, 151
Bone tracers, 31, 96, 131, 133
Boron (B), 33
Bremsstrahlung, 139, 140

C

Cadmium (Cd), 33, 82
Calcification, 17, 26, 108, 147
cartilage, 19
provisional, 20
Calcitonin, 122, 123, 125
Calcium (Ca), 13, 27, 31, 36, 37, 41, 105, 171, 195
absorption, 112-4, 119, 128
intestinal, 112, 120, 126, 127, 128
net, 113
regulation of, 113
balance, 113, 195
of body fluids, 106-8
$Ca^{40}$, 7, 11
$Ca^{45}$, 11, 22, 34, 55, 79, 92, 93, 96, 98, 111, 112, 116, 126, 127, 131, 134, 136, 156, 168
excretion, differential, 114
optical densities, 137
removal of, 95
uptake of, 91, 94, 95, 110
$Ca^{47}$, 11, 34, 36, 96, 98, 190, 192
external counting of, 140
gamma-emitter, 79, 189, 193
retention, 143
skeletal uptake of, 110
whole-body counts, 143
Ca/P ratio, 64, 65, 66, 67, 196
Ca/Sr discrimination, 116
carbonate, 67
concentration, homeostatic regulation of, 113
deposition, 91
endogenous, 113
estimates of annual intake (Table), 180
exchangeable, 95, 125, 189
excretion, 112-4
exogenous, 113
extraskeletal, exchangeable, 95
intake, 128
intraskeletal, exchangeable, 95
ionized, 105
labile pool of, 125
level, regulation of, 106
loss, 113-4
metabolism, kinetics of, 96
mobilization, 127
in normal serum, 107
proteinate, 107
retention, 91
secretion
differential, 114, 116
from blood to milk, 116
skeletal, quantities exchanged, 91
Sr/Ca, in human milk, 116
stable, in human milk, 116
transfer, 128
transport, 112
Calcium fraction
diffusible level, 106
ionized, 107
nondiffusible, 107
un-ionized, 107
Calcium ion, 27, 61, 63, 66, 67, 74, 111, 112, 151
concentration in blood, 122
homeostasis in fluids of body, 119-20
homeostatic regulation of, 123
mobilization, 122
nomogram for, 106, 107
Calcium phosphates, synthetic apatite, 64-5
Calcium pool, 125
miscible, see Miscible calcium pool (E)

Canalicules, 15, 21
  of haversian system, 22
  interconnecting, 18
Canaliculi, 24
Carbon (C), 11, 171
  $C^{12}$, 7
  $C^{14}$, 34, 39, 55, 79, 172, 176, 179, 183
Carbonate, 13, 61, 66, 74
  of bone, 67, 68-9, 79
  fluorapatite, 68
  hydroxyapatite, 68, 71
Carbonic anhydrase, 82
Cartilage, 18, 115
  calcification, 19
  calcified, 19
  chondrogenesis, 138, 155
  epiphyseal, 153
  hypertrophic, 19, 20
Cations, 66
  blood plasma, 119
  divalent, 63
Cell population kinetics, 155
Cells
  lining bone surfaces, 184
  mesenchymal, 28
  reticular, 15
Cement substance, 18
Cementoblasts, 28
Cementocytes, 24, 28
Cementoid, 24
Cementum, 23, 25, 26, 28
  primary, 28
  secondary, 28
Cerium (Ce), 151, 183
  $Ce^{133}$, 9
  $Ce^{140}$, 39
  $Ce^{144}$, 151, 169, 178
Cesium (Cs)
  $Cs^{137}$, 36, 45, 142, 178, 179, 183
Chelating agents, 183
Chemisorption, 75
Chondroitin sulfate, 126, 153
  turnover rates of, 156
Chromium (Cr), 33, 82
Citrate, 13, 61, 126
  of bone, 69
  as carrier, 111
Cobalt (Co), 33, 82, 85
Collagen, 14, 15, 16, 24, 69
  fibers, 13, 15, 28
    mineralization, 147
  fibrils, 18, 62, 72
    cross-banding, 15, 62, 69, 72
    fibrous long spacing (FLS), 16
    segment long spacing (SLS), 16
  native, 16
  reconstituted, 72
Compact bone, 21, 81, 133, 151
  remodeling, 134
  unit of structure, 21
Compartmental analysis, 99-100
Complexing agents, 183
Components of mineralized tissues, 24
  extracellular, 15-16
  intracellular, 15-16
Compton effect, 51, 52
Constituents of bone, 105-6
Copper (Cu), 33, 85, 86, 88
  deficiency, 88
Cortical bone, 195
  composition (Table), 67
  elementary analysis, 66
Cosmic rays, 171, 173, 176
Counting
  external, $Ca^{47}$ and $Sr^{85}$, 140
  grain, 136, 138
  track, 136, 138
  whole-body, 98-9, 131, 141-4, 190
Crossover time, 100
Crystals, see also Apatite crystals
  bone mineral, 17, 62-71, 168
    composition of, 72
  colloidal dimensions, 61
  discontinuities in, 65-6
  dynamics, 75
  growth, 17, 72, 74, 147
  hydroxyapatite, internal structure of, 65
  inorganic, 24
  lattice, 65
  nucleation, 15, 82, 147
  orientation, 74
  seeding, 82
  shape, 62
  size, 61, 62
  structure, 65
    internal, 63-4, 72, 74
    labile, 74

submicroscopic, 61
substitutions, 63
surfaces
chemistry, 74-5
ion exchange, 74, 75
ions, 74-5
labile structure of, 74-5
pH at, 74, 75
special conditions at, 75
unit cell, 63, 65
*a* axis, 65
*c* axis, 65
Curie (c), 159
Cushing's disease, 188
Cybernetic system theory, 120
control signal transducers, 121, 122, 124
derivative control, 121, 122
directed correction, 121
error sensing device, 120, 122, 123
integral control, 121, 122
proportional control, 121, 122
regulation of calcium ion concentration, 122
Cystine, $S^{35}$-labeled, 155

## D

Dahllite, 68, 71
Densitometry, 136
Dental tissues, 26-8
Dentin, 23, 24, 25, 26, 27, 66, 69-70
incremental lines of Ebner, 24
matrix, 69
mineral of, 69
secondary, 27
tubules, 24
Deoxynucleic acid (DNA), 155, 156
Deuterium ($H^2$), 10, 11
Deuteron induced transmutation, 10
Dial painters, 1, 167
Diethylenetriamine pentaacetic acid (DTPA), 183
2,4-Dinitrophenol, 111
Diphenyloxazole (phosphor), 53; (Table), 54
Discrimination
Ca/Sr, 116
calcium, 110, 112
Discrimination factor (DF), 109; (Table), 110
blood to urine, 114
soil to milk, 110
Dose commitment from nuclear testing, 183; (Table), 184
Dose rates, body tissue, 176; (Table), 176
Dual mechanism, 123-4
Dwarfism, 84

## E

Electromagnetic radiation, 8, 47, 51
Electron capture (EC), 9, 33, 162-5
transmutation by, 38
Electronic analog computer, 96, 99, 140
Electrons, 3, 8, 47, 52, 139, 140, 171
Elements, 7; (Table), 4-6
bone-seeking, 31-46, 91
heavy, 10, 33, 171
translocation, 39-41
transmutation, 39-41
periodic table, 31-3
relative atomic mass, 7
transuranic, 33
Enamel, 23, 24, 25, 26-7, 66, 70-1
calcified, 27
contour lines, 27
epithelial origin of, 71
epithelium, 28
formation, 27
lines of Retzius, 24, 27
matrix, 71
mineral of, 70
morphology of, 70
organic content of, 71
organs, 23
prism sheaths, 26, 27
prisms, 26
rods, 71
Endochondral ossification, 18
Endogenous factor, 113, 128
Epitaxy, 74
Ergocalciferol-$C^{14}$ (vitamin D), 157
Estrogens, 14, 156
Estradiol 17-β benzoate, 151
Estradiol-6,7 $H^3$, 132, 157
Estrone-16 $C^{14}$, 157
Ethylenediamine tetraacetic acid (EDTA), 81, 183
Eukeratin, 71
Europium (Er), 10, 151, 177
Excitation, 47, 48, 51
Excitation principle, use of, 51

Excretion
  calcium, 112-4
  $Sr^{85}$, cumulative, 99; (Table), 99
  strontium, preferential, 114
Exponential functions, 100, 101, 102
Exostoses, 131

F

Fallout, 39, 110, 182
  local, 178
  radioactive, 177-9
  tropospheric, 178
Feedback, 123
  mechanism, 124
  negative, 120
  transducer, 120-1, 122, 123
Fibers
  bone, 18
  elastic, 28
  reticular, 13, 25
  Sharpey's, 28
Film badge, 59
Fission products in bone, retention (Table), 178
Fluids of body
  calcium of, 106-8
  intercellular, 105, 106, 108
  phosphates of, 108-9
Fluorapatite, 84, 85
  carbonate, 68
Fluorescence, 133, 134
Fluoride (F), 84-5
  $F^{18}$, 9, 34, 55, 85, 140
  ion, 61, 63, 67, 68
Fluorosis, 85
Food chain, importance of milk in, 110, 114
Fractures, 97, 190, 192
  bone lesions in, 140
  callus, 97
  external counting of $Ca^{47}$ and $Sr^{85}$, 140
  healing of, 13-14
  $S^{35}$ in repair, 153
Francolite, 68

G

Gallium (Ga), 33
  $Ga^{72}$, 140
Gamma rays, 8, 9, 36, 37, 49, 51, 52, 53, 58, 88, 98
  penetrating, 177
  primary, 139
  spectrometry, low level, 144
  spectrum, 142
Gamma scanner, high-energy (HEG), 140, 192, 193
Gas amplification, 50
Geiger-Müller (G-M) tube counters, 48, 50, 55, 58, 136, 140, 141, 142
Geiger region, 50
Genetic mutations, 39
Glucagon, 123
Glycine, 15
  $C^{14}$-labeled, 155, 156
  tritiated, 155, 156
Glycolysis, intracellular, 111
Glycolytic cycle, 124, 125
Glycoprotein, 16, 26, 27
Grain counting, 136, 138
Ground substance, 13, 15-16, 18, 26, 29, 31
Growth apparatus, 82
Growth-arrest lines, 131

H

Half-life
  biological, 39, 161
  physical, 39, 161
  of radioisotopes, 9
Haversian systems (osteons), 18, 82, 87
  canalicules, 22
  canals, 22, 28, 155
  cementing line of v. Ebner, 22
  primitive, 21
  unit of structure of compact bone, 21
Hexosamines, 15, 16
Histiocytes, 149
Histogenesis of teeth, 23, 25-6
Homeostasis of calcium ion
  cybernetic aspects of, 120-2
  dual mechanism, 123
  in fluids of body, 119-20
  in mineral metabolism, 119-28
  regulation of, 123
Homeostatic control, 121
Homeostatic mechanisms, 105
Hotspots, 22, 88, 135, 136
Hydration shell, 63
Hydrogen (H)
  $H^1$, 10, 11

$H^2$ (deuterium), 10, 11
$H^3$ (tritium), 11, 34, 55, 172
bonds, 64, 67
Hydroxyproline, 15, 156
Hyperparathyroidism, 189
Hyperthyroidism, 189
Hypervitaminosis D, 127
Hypogonadism, 84, 188
Hypoparathyroidism, 127, 189

## I

Information theory, 120
Insulin, 123
Integral spectrum, 58
Integrated serum activity curve, 93
Intercellular substance, 13, 23
Internal environment, 105-6
International atomic weights (Table), 4-6
Interprismatic substance, 26
Interstitial fluids, 15
Interstitial lamellae, 21
Interstitial substance, 13, 15-17, 22
Intracartilaginous ossification, 18
Intramembranous ossification, 18
Iodine (I), 33, 85
$I^{131}$, 178, 179
Ion collection, 48
Ion exchange, 68, 74, 75, 85, 93, 98, 106, 114, 168, 182
heteroionic, 135
isoionic, 135
long-term, 41, 136
Ionization, 47, 48, 51
by excitation, 52
secondary, 50
specific, 48-9, 159
Ionization chambers, 48-50, 58, 141, 142
Ionization principle, use of, 48-9
Ion pump, 111
Ion transport
across biological membranes, 111-2
net, 112
Iron (Fe), 33, 85, 171
Isomeric transition (IT), 9
Isomorphic substitution, 64

## K

Keratin, 24
Kinetics
of bone mineral, 91-102
cell population, 155
of decay of radioelements, 100
methods, 187-8, 191
studies, 195
interpretation of, 92-8
results of, 189
of strontium metabolism, 140

## L

Labeled organic molecules, 153-7
biosynthesis, 153
precursors, 153
turnover rates, 155
Lactic acid, 124
Lactose, 112
Lacunae, 15, 18, 21, 22, 24
Lamellae, 21
circumferential, 21
fibrillated, 22
interstitial, 21
new, 134
Lanthanide series, 33
Lanthanons, 151
absorption from intestinal tract, 151
Lanthanum (La), 33, 39, 151, 178
$La^{138}$, 172
$La^{140}$, 35, 39, 178
Lead (Pb), 33, 40, 54, 63, 131
$Pb^{208}$, 44
$Pb^{210}$, 176
Leucine
$C^{14}$-labeled, 155, 156
tritiated, 155
Linear energy transfer (LET), 160
Lithium (Li)
$Li^7$, 9, 55
Lutetium (Lu), 151
$Lu^{176}$, 172
Lysine, 112

## M

Madder, 131
Manganese (Mn), 33, 82, 85, 86, 88
$Mn^{54}$, 34
Magnesium (Mg), 31, 126
in bone, 81
deficiency, 81
ion, 61, 67, 111, 112
$Mg^{28}$, 34, 55, 81

Markers
  chemical, 131
  metal, 131
Marshall Islanders, 178
Mass law equation, 107
Mass (A) number, 6
Mast cells, 14, 29
Matrix, 15, 27, 31, 39
  collagenous, 27
  extracellular, 18
  precursor, 156
Maximum permissible body burdens (MPBB), 166; (Table), 162-5
Maximum permissible concentrations (MPC), 166
Measurement of radioactivity, 47-60; (Table), 54-7
Mesons, 171
Mesothorium$_1$ ($Ra^{228}$), 1, 44, 149, 167, 168; (Table), 43
  in dial painting, 167
  transmutation of, 149
Mesothorium$_2$ ($Ac^{228}$), 43, 44, 149
Metachromasia, 23
Metachromatic ground substance, 29
Metal-enzyme complexes, 82
Methionine
  $S^{35}$-labeled, 155
  tritiated, 155
Microautoradiography, 155
Microcurie (μc), 159
Micro-microcurie (μμc), 159
Microradiography, 22, 82, 83, 87
Microsomes, 88
Milk, $Sr^{90}$ in, 181-3; (Table), 180
Millicurie (mc), 159
Mineral metabolism, 105-16
  homeostasis, 119-28
Mineralization
  bone, 82, 147
  inhibitor of, 82
Miscible calcium pool (E), 95, 97, 187, 190
Molybdenum (Mo), 33, 85, 88
Monovalent anions, 63
Mucopolysaccharides, 13, 69
  sulfated, 31, 72, 125-6
Multichannel pulse-height analysis, 98
Mutations, genetic, 39

## N

Negatron (β−), 8, 34, 35, 37, 38, 42, 43, 55, 162-5
Neon ($Ne^{22}$), 9
Neonatal line, 27
Neptunium series: $Np^{237}$ to $Bi^{209}$, 40
Neutrino, 8
Neutrons, 3, 6, 10, 39, 171
Nickel (Ni), 33
Niobium (Nb), 33, 178
  $Nb^{95}$, 35
Nitrogen (N), 11, 171
  $N^{14}$, 39
Nuclear fission, 177
Nuclear species, 7
Nuclear structure, 2-7
Nucleation of crystals, 15, 22, 82, 147
Nucleic acid, 37
Nucleonic components, 171
Nucleons, 3
Nuclide, 6
  radioactive, 10-11
  relative nuclidic mass, 7

## O

Observed ratio (OR), 109
  data (Table), 110
  values, Sr/Ca, 110, 115
Octocalcium phosphate, 65
Odontoblasts, 25, 26, 28
Organic matrix, 13, 17, 18, 22, 40, 44, 88, 147
  depolymerization, 124
  solubilization, 124
Organic molecules, labeled, 153-7
Organic portion of bone, 147-57
Organization of bone, 18-22
Orthophosphate, 37, 65
Orthophosphoric acid, 108
  dissociation constants (Table), 108
Ossification, 24
  endochondral, 18
  intracartilaginous, 18
  intramembranous, 18
  primary, 18
Osteitis deformans, 191-2
Osteoblasts, 15, 18, 21, 122, 124, 152, 153, 155

function of, 13
labeled, 152, 156
location of, 14
radiation-damaged, 166
Osteoclasts, 14, 88, 122, 124, 152, 153, 155
activity, 125, 151
autoradiograph, 148, 152
function of, 13, 15, 16, 125
radiation-damaged, 166
Osteocytes, 14, 15, 21, 24, 122, 152, 153, 155, 166
autoradiograph, 152
death of, 149
function of, 13
Osteogenesis, 138
Osteoid, 24
rachitic, 73
uncalcified, 151
Osteomyelitis, 190
in dial painters, 1
Osteons, 21, 22, 81, 125
compact bone, 134
lead in, 131
secondary, 21
Osteopenia, 188, 189, 190, 193-6
Osteoporosis, 98, 189, 190, 193-6
in poliomyelitis, 195
postmenopausal, 195
Osteosclerosis, 188
Oxidative phosphorylation, 111
Oxygen (O), 171
$O^{18}$, 9


## P


Paget's disease, 140, 188, 189, 190, 191-2
Parathyroid activity, 123
unifying concept of, 124-5
Parathyroid adenoma, 188
Parathyroid extract, 14, 123, 124, 125, 126, 127
Parathyroid glands, 112, 120, 122, 123, 124, 125
Parathyroid hormone, 69, 119, 122-3, 127
isotopes used in study of, 125-6
mode of action on bone, 124-5
Periodic table of elements, 31-3
Periodontal membrane, 23, 25, 26, 28, 29
Periosteum, 29
Phenylalanine-$C^{14}$, 156
Phosphate, 13
absorption, 119
ions, 27
inorganic, 105-6, 108
transport of, 108, 125
interchange of, 91
Phosphors, 51, 53, 58, *see also* Scintillators
physical constants, 54
Phosphorus (P), 31, 37, 195
$P^{32}$, 22, 37, 55, 72, 78, 131, 134, 140
beta emitter, 166
in rickets, 196-7
tumors caused by, 32
Photoelectric effects, 51, 52
Photomultiplier tube, 53
Photons, 8, 51, 171
Picocurie (pc), 159
Placental discrimination, 110
Plutonium (Pu)
$Pu^{239}$, 88, 147-9, 151
in bone, 88
removal from blood, 149
toxicity of, 167
valence states, 149
Polonium (Po)
$Po^{214}$, 41
Polysaccharides, 24
sulfated, 15-16
Positron ($\beta+$), 8, 33, 34, 35, 52; (Table), 37, 38
Positron annihilation, 139, 140
Potassium (K), 31, 36, 37, 111
$K^{40}$, 142, 172, 173, 176
Power function, 100-2
Praseodymium (Pr), 151, 178
Prechondroblasts, 155
Predentin, 24, 26
Preodontoblasts, 155
Preosseous tissue, 82, 83, 156
Preosteoblasts, 155
Proline, $C^{14}$-labeled, 154, 155, 156
Promethium (Pm), 151, 178
Proportional counters, 48, 55, 58
Proportional region, 50
Protein-polysaccharides, 156
Protons, 3, 6, 171
transmutation induced by, 10
Pulse-height analysis, multichannel, 98
Pyrophosphate, 65, 66, 67

Pyrrolidine amino acids, 15
Pyruvic acid, 124

## R

Rad, 159, 160
Radiation
  and bone tumors, 167-9
  effects upon bones, 159-69
  electromagnetic, 47, 48
  external, 160-1; (Table), 174
  geomagnetically trapped, 171
  internal, 161, 166-7
  ionizing, 1, 173
  natural, 173
    cosmic, 171, 173, 176
    galactic, 171
    solar, 171
  secondary, 171
  terrestrial, 176
  units, 159-60
  whole-body exposure to, 183-4
Radiation detectors, 65
  associated equipment, 56-7
  current-producing system, 59
  nonpulse systems, 48, 59-60
  nuclear, 48
  photographic system, 59
  pulse systems, 48-58
Radioactivity, 7-9, 47-60
  detection, 47-60
  external counting, 79
  measurement, 47-60; (Table), 54-7
  natural, 171-2, 173, 176
    body tissue dose rates, 176, 196
  in sea water, 172
  $Sr^{90}$, primary sources of, 184
  units, 159-60
  in water, 172
Radioassay, 55, 131, 137-9
Radioelements (radioisotopes), 10, 39, 71, 147-57
  in bone, 79-88, 135, 161, 166
  deposition
    diffuse component, 88
    in osteons, 135
  distinguished from trace elements, 85-8
  in fallout, 177
  hotspots, 22, 88, 135, 136
  osteogram, 194, 195
  removal from bone, 183
    carrier agents, 183
  scanning of bone, 139-40
  special interest for bone, 34-5
  as tracers, 2
Radionuclides, 10, 33, 172
Radiothorium ($Th^{228}$), 40, 41, 44, 149-50, 167, 168
Radium (Ra), 1, 31, 98, 168, 183
  body burden, 167
  decay products, 42
  dial painters, 1, 167
  emanation, 40
  heavy element, 10
  maximum permissible concentrations (MPC), 166
  poisoning, 88
  $Ra^{224}$ (Thorium X), 40, 44, 151
  $Ra^{226}$, 11, 35, 40, 41, 42, 101, 133, 151, 168, 172, 173, 176
    total body content, 175, 176
  $Ra^{228}$ ($Mesothorium_1$), 40, 43, 44, 45, 149, 167, 168, 176
Radium labeled system, 87
Radon (Rn, radium emanation), 40
  $Rn^{220}$ (Thoron), 40, 44
  $Rn^{222}$, 40, 41, 172, 176
Recrystallization, 61, 65
Recycling, 128
Relative biological effectiveness (RBE), 160, 166
  of alpha rays, 160
  alpha to beta radiation, 161
  concept of, 161
  values, 159
Renal tubular defects, 188
Resorption, 13, 18, 19, 21, 92, 93, 94, 98, 119, 127, 193-4
  osteoclastic, 122, 124, 125
  reactions, 95
  of stable bone mineral, 125
Resorption cavities (absorption cavities), 21, 166
Retention, total body, 96
Rickets, 113
  vitamin D-deficient, 189, 190, 196
  vitamin D-resistant, 187, 190, 196-7
Roentgen equivalent man (rem), 160

Roentgen unit (r), 159
Rubidium ($Rb^{85}$), 9, 38

## S

Samarium ($Sm^{147}$), 172
Sarcoma, osteogenic, 1, 45, 167, 168, 169
Scandium (Sc), 37
  $Sc^{47}$, 36
Scanning, skeletal, 140
Scintillation, 51
  counters
    collimated, 96, 139, 140
    well-type, 55
  crystal, 140
  detectors, 48, 51, 53, 58
  phosphors, 88
    liquid, 54, 55
    physical constants, 54
    plastic, 53, 54
Secretion
  of dietary $Sr^{90}$, 116
  of Sr from blood to milk, 116
Sharpey's fibers, 28
Silicon (Si), 33, 37
  ion, 63
Silver (Ag), 33
Skeletal abnormalities, 160
SLS, segment long spacing, 16
Sodium (Na), 31, 61, 79, 80, 111
  of bone, 79-81
  exchangeable, 80, 91
  ion, 61
  mineral phase of, 80
  nonexchangeable, 80
  $Na^{22}$, 9, 34, 79, 80, 81, 91
  $Na^{23}$, 80
  $Na^{24}$, 55, 79, 80
  NaI(Tl) (phosphor), 51, 53, 54
Soil-food chain, 173
Solubility relationships, 71-2
Solubilization
  of bone mineral, 74
  of organic matrix, 124
Specific activities
  of blood, 92, 101, 188
  of bone, 91, 181
  concept of, 139
  curves, 101
  of whole body, 101
Spectrometry
  infrared, 68
  low-level, gamma ray, 144
Spectrum
  gamma ray, 142
  integral, 58
Staffelite, 68
Steatorrhea with osteomalacia, 188
Stilbene (phosphor), 54
Strontium (Sr), 31, 33, 63, 85, 167, 168, 179, 183
  discrimination, 110, 112
  exchangeable, 96, 100
  excretion, preferential, 114
  ion, 111, 112
  metabolism
    kinetics of, 95, 140
    model for, 96
  secretion, blood to milk, 116
  $Sr^{85}$, 9, 11, 34, 96, 112, 114, 116, 190, 192, 194
    excretion, cumulative, 99
    external counting of, 140
    gamma emitter, 79, 99, 189, 193
    skeletal uptake of, 110
  $Sr^{89}$, 11, 34, 114, 127, 169, 178, 183
  $Sr^{90}$, 11, 34, 37, 38, 110, 116, 151, 166, 167, 169, 178, 179-83
    beta emitter, 166
    in bone, 182
    dietary, secretion of, 181-2
    distribution, in autoradiographs, 134-5
    estimates, annual intake, 180
    exposure to, 183
    in milk, 180, 181, 182-3
    Sr/Ca, 110, 116, 180, 181
    source of radioactivity, 184
Sulfate ion, 153
Sulfated mucopolysaccharides, 31, 72, 125-6
Sulfur (S), 31, 37
  $S^{35}$, 31, 34, 55, 125, 126, 151, 153, 156
    labeled amino acids, 155
    labeled precursors, 155

## T

Thorium (Th), 43, 149, 151, 176, 183
  $Th^{228}$ (Radiothorium), 40, 41, 44, 149-50, 167, 168

- $Th^{230}$, 40
- $Th^{232}$, 40, 41, 44, 45, 151, 167, 172, 173
- $Th^{234}$, 35, 40

Thorium series: $Th^{232}$ to $Pb^{208}$, 40, 41, 43-5, 176
Thorium X ($Ra^{224}$), 44
Thoron ($Rn^{220}$), 44
Thorotrast ($Th^{232}$), 41, 43, 44, 45, 167
Thymidine, 156
- $C^{14}$-labeled, 155
- tritiated, 138, 155

Tin (Sn), 33
Titanium (Ti), 33
- $Ti^{47}$, 36

Tooth germs, 23
Trace elements, 33, 85-8
Tracer elements, 2, 11, 31, 108, 196-7
Tracer techniques for bone, 131-44
Transport
- of calcium, 112, 128
- of mineral, bone to blood, 74
- of phosphate, 108, 125

Transformations, cellular, 13, 14
Translocation, 33-45
- in heavy elements, 39-41

Transmutation, 33-45
- alpha particle induced, 10
- deuteron induced, 10
- by electron capture, 38
- in heavy elements, 39-41
- proton induced, 10
- of $Ra^{228}$ (mesothorium$_1$), 149
- relationship to translocation, 40
- with translocation, 39

Tricalcium phosphate
- alpha, 64
- beta, 65
- hydrated, 64

Tricarboxylic acid cycle intermediates, 155
Tritiated amino acids, 155-6
Tritiated estrogens, 132
Tritiated thymidine, 138, 155
Tritium ($H^3$), 11, 19, 176
Tropocollagen, 16, 156
Tumors, 45
- body burden of radium, 167
- bone, 45, 140, 167-9, 188, 189, 190
- caused by $P^{32}$, 169
- hepatic, 45
- induction period, 167
- malignant, bone, 167
- metastases, 140, 188, 190
- periosteal, 45


## U


Ultraviolet light, 47
Uranium (U)
- $U^{233}$, 35
- $U^{238}$, 41, 172, 173

Uranium-actinium series: $U^{235}$ to $Pb^{207}$, 40
Uranium-radium series: $U^{238}$ to $Pb^{206}$, 40, 41, 42, 176


## V


Vascular channels, 149
Vitamin D, 69, 111, 112, 113, 114, 120, 126-8, 157
- calcemic effect, 127
- on calcium absorption, 128
- deficient rickets, 189, 190, 196
- endogenous, 119, 126
- exogenous, 119, 126
- resistant rickets, 189, 190, 196-7


## W


Water, 16-17
- of bone, 17
- radioactivity in, 172

Whole-body counting, 98-9, 141-4, 190
- development, 141
- external scanning, 131

Whole-body exposure to radiation, 183-4
Whole body, specific activities of, 101


## X


X-ray diffraction, 15, 23, 62, 63, 64
X-rays, 8, 49, 51


## Y


Ytterbium (Yb), 151
Yttrium (Y), 33, 37, 38, 125, 151, 153, 169, 178, 183
- $Y^{90}$, 35, 37, 38, 55, 151, 168, 169, 178
- $Y^{91}$, 35, 151


## Z


Zinc (Zn), 10, 33, 82, 84, 85, 177
- in bone, 81-2, 84

deficiency, 82, 84
exchangeable, 82
$Zn^{65}$, 34, 82
Zirconium (Zr), 33, 35, 38, 178, 183
$Zr^{90}$, 38, 178
$Zr^{95}$, 35
ZnS(Ag) (phosphor), 54
Z number of elements, 4-6, 8, 140